£8·75p

Infinitesimal Calculus

Infinitesimal Calculus

James M. Henle

Eugene M. Kleinberg

The MIT Press
Cambridge, Massachusetts, and London, England

This book was printed and
bound by Alpine Press in the
United States of America.

**Library of Congress Cataloging in
Publication Data**

Henle, James M
  Infinitesimal calculus.

  Includes index.
  1. Calculus. I. Kleinberg,
Eugene M., joint author. II. Title.
QA303.H486   515   78–18849
ISBN 0–262–08097–4

to our parents

. . . we shall discover much Emptiness, Darkness and Confusion; nay, if I mistake not, direct impossibilities and contradictions. Whether this be the case or no, every thinking reader is entreated to examine and judge for himself . . .

George Berkeley, Bishop of Cloyne (1685–1753)

Contents

Preface

Calculus has always been a difficult subject to learn well. In the last half century alone there have been literally scores of calculus books published, each trying harder than the next to simplify the subject.

At the turn of the century it was popular to teach calculus by using so-called infinitesimals. This approach had the advantage of making the basic theory quite intuitive and easy to understand, but mathematicians lacked a rigorous definition of just what infinitesimals were, and so anyone advancing beyond the basics quickly became lost. Thus we were led to the ε/δ approach to calculus, an approach that, although totally precise and rigorous, was a disaster for students to learn and teachers to teach. Most recently ε's and δ's have been shelved along with all other attempts at teaching the basics of calculus. Instead we have settled into teaching specific methods for applying calculus in specific situations. The problem here, of course, is that even though individual methods might be fairly easy to master, there exist very many, seemingly distinct, methods to be learned, and even then most courses leave students hopelessly short.

This active evolution in the teaching of calculus was always prompted by continual dissatisfaction with earlier approaches and had nothing to do with new mathematical insights into calculus itself. Indeed there *were* no such "new insights"; that is, there were none until just recently.

In the early 1960s the mathematician Abraham Robinson pioneered a body of work known as Nonstandard Analysis which makes precise and mathematically rigorous the intuitively pleasing concept of infinitesimal. Originally Robinson's field was reserved as an advanced graduate subject, but the ideas are both simple and important, and in recent years everyone from standard mathematicians to economists, physicists, and social scientists have been using his methods with stunning success.

A most natural place for Robinson's insight is as a next (and possibly final) point in the evolution of the teaching of calculus. We can now develop calculus using infinitesimals and enjoy all of their simplicity and intuitive power, yet at the same time work in a mathematically precise and rigorous atmosphere. This approach, although quite new, has been used at a number of universities with remarkable success.

This book presents a rigorous development of calculus using infinitesimals in the style of Robinson. It does not make any attempt to cover the assorted methods of calculus for applications, but rather it concentrates on theory, the area which previously was so difficult. We feel that with this new approach, basic theory is now quite accessible to students—even those who are interested in calculus solely for its applications. Indeed a knowledge of basic theory lets one dispense with learning many of the canned methods in favor of attacking problems directly and formulating one's own methods.

The only prerequisite assumed for this book is a good foundation in high school mathematics.

Our early chapters deal with the field of mathematical logic. Some of this is necessary for an understanding of infinitesimal calculus, but much of it is not. That which can be skipped is indicated in the text.

We received help from a great many people during the preparation of this manuscript, much of it from students taking preliminary versions of the course. Special thanks are also due to Professors Frank Wattenberg and David Schaffer, and to Mitchell Spector. Finally, Denise Borsuk typed countless versions of the text and displayed patience, good humor, and skill. To her we are most grateful.

J. M. Henle
E. M. Kleinberg

Infinitesimal Calculus

1

Introduction

The history of modern mathematics is to an astonishing degree the history of the calculus. The calculus was the first great achievement of mathematics since the Greeks and it dominated mathematical exploration for centuries. The questions it answered and the questions it raised lay at the heart of man's understanding of not only geometry and number, but also space and time and mathematical truth. It began with the surprising unification of two rather different geometrical problems, and almost immediately its ideas bore fruit in dozens of seemingly unrelated areas. The methods it developed gave the physical sciences an impetus without parallel in history, for through them natural science was born, and without them physics could not have progressed much further than the mystical vortices of Descartes.

In the beginning there were two calculi, the differential and the integral. The first had been developed to determine the slopes of tangents to certain curves, the second to determine the areas of certain regions bounded by curves. Algebra, geometry, and trigonometry were simply insufficient to solve general problems of this sort, and prior to the late seventeenth century mathematicians could at best handle only special cases.

The general idea of the calculus, its fundamental theorem, and its first applications to the outstanding problems of mathematics and the natural sciences are due independently to Isaac Newton (1642–1727) and Gottfried Leibniz (1646–1716).

Their work was certainly built on foundations laid by others, but their penetrating insights represented what is easily the most significant mathematical breakthrough since the Greeks. Remarkably, the powerful

Think, for example, about how one unfamiliar with calculus would go about finding the slope of the tangent to the curve

$$y = x \sin \frac{\pi}{2x}$$

at (1, 1):

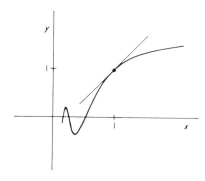

Or how about the problem of finding the area of the plane region bounded by the curve $y = x^2$ and the lines $y = 0, x = 1,$ and $x = 2$:

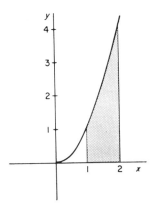

These problems are child's play for any student armed with elementary calculus.

The greatest contribution of Newton and Leibniz lies in their abstraction, organization, and notation. While each made his discoveries independently, there were later accusations that Leibniz's work was not original. The debate was frequently bitter and the chief result was that English mathematicians patriotically refused to use Leibniz's far superior notation. This was possibly responsible, in large measure, for England's loss of scientific leadership during the Enlightenment.

Ironically, while supporters of Newton and Leibniz argued bitterly over the discovery of the calculus, neither side knew that a third man had made the discovery independently, Seki Kōwa (1642–1708) in Japan. The *yenri*, or "circle method," an early form of the calculus, was found at about the same time as Newton and Leibniz, and, although there is no direct evidence, it is most likely that this was due to Seki Kōwa.

methods developed by these two men solved the same class of problems and proved many of the same theorems yet were based on different theories. Newton thought in terms of limits whereas Leibniz thought in terms of infinitesimals, and although Newton's theory was formalized long before Leibniz's, it is far easier to work with Leibniz's techniques.

The approach to the calculus we shall employ is based on Leibniz's ideas as formalized by Abraham Robinson in 1961 under the name of "nonstandard analysis." Simply stated, our approach will involve *expanding the real number system* by introducing *new* numbers called "infinitesimals."

These new numbers will have the property that although different from 0, each is smaller than every positive real number and larger than every negative real number. Of course our infinitesimals cannot themselves be real numbers, but so what? This sort of expansion of a number system through the introduction of new numbers which themselves correspond to nothing in the real world is common in mathematics. Negative numbers and imaginary numbers have no direct physical presence *in* the real world, yet both serve an essential role in solving problems *about* the real world. These new infinitesimals, once suitably defined, will enable us to solve general problems of slopes of tangents and areas of regions with extraordinary ease. Here's an example:

We shall find the slope of the tangent to $y = x^2$ at the point $(1, 1)$.

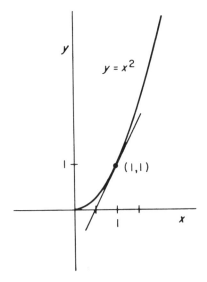

How can we approach this problem? We know how to find the slope of a line given two points on the line, but here we are given only one point plus the information that the line is tangent to $y = x^2$ at that point. Our solution is simple: We let ◎ be an infinitesimal (positive, say) and consider two points on the curve of $y = x^2$ which are *infinitely close* to one another, $(1, 1)$ and $(1 + ◎, (1 + ◎)^2)$:

Keep in mind that since we haven't formally defined infinitesimals or discussed their properties, this example takes steps which are as yet unjustified.

The symbol ◎ may be pronounced "hype."

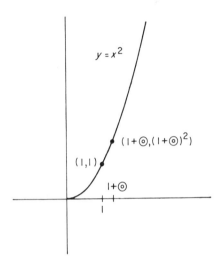

We can certainly find the slope of the line going through these two points:

Later in his life, Newton became quite suspicious and guarded—at the encouragement of friends. Fearing that Leibniz might steal his ideas, he was once deliberately obscure. One letter to Leibniz reads in its entirety: "*6a, 2c, d, ae, 13e, 2f, 7i, 3l, 9n, 40, 4q, 2r, 4s, 9t, 12v, x.*"

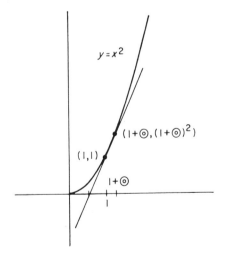

It is

$$\frac{\varDelta y}{\varDelta x} = \frac{(1 + ◎)^2 - 1}{(1 + ◎) - 1} = \frac{◎^2 + 2◎}{◎} = 2 + ◎.$$

As originally formulated by either Leibniz or Newton, the calculus left much to be desired. It was vaguely presented, lacked a solid foundation, and was used often by people who could not satisfactorily explain it. As such, it was subjected to many attacks. One of the most famous was Bishop Berkeley's *The Analyst, or A Discourse Addressed to an Infidel Mathematician* from which the quotation on page vi is taken. Berkeley's attack was a serious one which was not adequately answered for over one hundred years. The "Infidel Mathematician," incidentally, was Edmund Halley, a friend of Newton's, who had earned the wrath of Berkeley by converting a mutual acquaintance to atheism by "mathematical" reasoning.

Similarly, the slope of the chord going through (1, 1) and $(1 - \odot, (1 - \odot)^2)$ is $2 - \odot$.

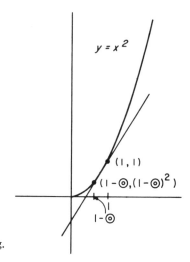

So, what is the slope of our desired tangent? Well, the slope of the tangent must be a *real* number ($2 + \odot$ and $2 - \odot$ are not real numbers), and it must fall between the slope of the chords BA and AC.

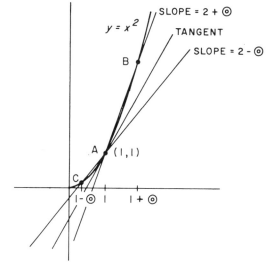

For contrast, let's look at a Newton-style proof that the slope is 2: For any Δx the ratio $\Delta y/\Delta x$ is an approximation of the true slope of the curve.

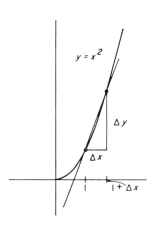

Two is such a real number, and it is easy to see that 2 is the only such number: Since $0 < \odot < r$ for every real $r > 0$, there can be no reals between 2 and $2 + \odot$, and similarly, since $-r < -\odot < 0$ for every real $r > 0$, there can be no reals between $2 - \odot$ and 2. Thus 2 is the only real number between $2 - \odot$ and $2 + \odot$, and so 2 must be our desired slope.

As Δx becomes smaller and smaller, the approximation improves, and the *limit*, as Δx approaches 0, is the desired slope. To be precise, if $f(x)$ is any function, we define *the limit of $f(x)$, as x approaches a, equals b*

$$\left(\lim_{x \to a} f(x) = b \right)$$

by "for all $\varepsilon > 0$, there is a $\delta > 0$ such that whenever $0 < |x - a| < \delta$, then $|f(x) - b| < \varepsilon$." To apply this, we first define a function $m(\Delta x) = \Delta y / \Delta x$. By calculations we find that, given Δx, Δy is

$2\Delta x + (\Delta x)^2$. Thus

$$m(\Delta x) = \frac{2\Delta x + (\Delta x)^2}{\Delta x}.$$

Finally we prove that according to the definition of limit,

$$\lim_{\Delta x \to 0} m(\Delta x) = 2,$$

that is, for every $\varepsilon > 0$, there is a $\delta > 0$ such that whenever

$$0 < |\Delta x - 0| < \delta,$$

then

$$\left| \frac{2\Delta x - (\Delta x)^2}{\Delta x} - 2 \right| < \varepsilon.$$

The proof goes as follows: Suppose $\varepsilon > 0$ is given. Then let (by guessing) δ equal ε. We can now check that this δ works: If $0 < |\Delta x - 0| < \delta$, then

$$\left| \frac{2\Delta x + (\Delta x)^2}{\Delta x} - 2 \right| = |2 + \Delta x - 2|$$

$$= |\Delta x|$$

$$= |\Delta x - 0|$$

$$< \delta = \varepsilon.$$

Notice that if Δx actually equals 0, then $\Delta y = 0$ too and $m(0) = 0/0$, which is undefined. The mystery is that although both the numerator Δy and the denominator Δx approach 0 as Δx approaches 0, the fraction itself approaches 2. This limit Newton called the "ultimate ratio."

In Newton's words: *"The ultimate ratios in which quantities vanish are not really the ratios of*

The expansion of number systems has occurred often in the history of mathematics and has usually marked a major turning point. Our problem of desiring a new number, an infinitesimal, is entirely similar to that, say, of the early algebraists struggling to solve equations. To these men the known number system consisted of 0, the positive rational numbers, and possibly a few irrationals like $\sqrt{2}$, $\sqrt{3}$, and π. When faced with an equation such as $2x + 10 = 6$, they recognized no possible solution. Later generations of mathematicians invented symbols to represent the solutions to such equations, solutions we would call negative numbers, but the inventors still denied their existence. They recognized them only as symbols which could be manipulated in equations, but which were not actually *numbers*.

The adoption of negative numbers was a slow process which was not completed for centuries. During this period they were tested thoroughly for consistency to see if they could be used without harm. Only when a number system was solidly constructed which contained both positive and negative numbers were they finally accepted. The most interesting point is that these numbers had a formal existence long before they became accepted. At first they were mere symbols, and then only after centuries of use did they become numbers.

A similar story can be told of the birth of complex numbers. As early as 1545, Cardan had formal symbols for them which he enjoyed manipulating, but which he regarded as fictitious. Soon everyone was using them, but again only in a formal way. The great mathematician of the eighteenth century, Leonhard Euler, remarked of them: " . . . and of such numbers we may truly assert that they are neither nothing, nor greater than nothing, which necessarily constitutes them imaginary or impossible. . . ." Recognition and acceptance of the complex numbers as numbers wasn't actually final until the nineteenth century.

The history of algebra is, in large part, the history of number. From the earliest conception of 1, 2, 3, ... our idea of number has grown slowly and painfully. At each step the growth was the result of a need. The need for numbers to represent the solution to equations such as $2x + 10 = 6$ led to the negative numbers. The need for numbers to represent $10 \div 4$ and $1 \div 3$ led to the

ultimate quantities, but the limits toward which the ratios of quantities, decreasing without limit, always approach; and to which they can come nearer than any given difference, but which they can never pass nor attain before the quantities are diminished indefinitely."

We should point out here that the example for our Newton-style proof is about as simple as they come. With more complex examples the choice of an appropriate δ given an ε becomes inordinately difficult. By contrast, proofs using infinitesimals become no more difficult. Fortunately, we do not plan to mention Newton-style proofs again in this text.

rational numbers. The need for numbers to represent the diagonal of a square of side 1

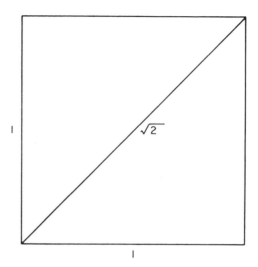

and the circumference of a circle of radius 1

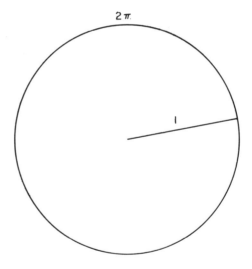

Amazingly, problems of both tangents and areas had actually been explored by the Greeks with remarkable success, considering the limitations of mathematics at the time. Archimedes, one of the greatest of the Greek mathematicians, was able to construct the tangent to the spiral that bears his name,

no easy task, even for a modern student equipped with a term of calculus. Even more amazing, Archimedes found the areas of many curved regions, including the parabolic segment:

led to the irrational numbers. The need for a root to the equation $x^2 + 1 = 0$ led to the complex numbers.

Many of the ideas of the calculus were present in the mind of this man, but without analytic geometry, without even an adequate system of numeration, he was unable to proceed further. His murder in 212 B.C. by a Roman soldier symbolized the end of creative mathematics for nearly 1,800 years.

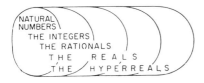

Oddly enough, Archimedes knew this too. Archimedes' proofs, as he published and communicated them to Conon, used the "method of exhaustion," a limit-type proof. For centuries this method was held as a model for calculus proofs. It was one of the chief reasons seventeenth- and eighteenth-century mathematicians rejected infinitesimals and eventually made Newton's approach rigorous. It was only early in the twentieth century that a letter of Archimedes' was found in which he expounded on his method, not of *proving* theorems,

Finally, the need for the calculus led to the infinitesimals. At each step expansion of the number system met with opposition, and at each step the new numbers were formally accepted long before they were given the status of numbers. As late as the 1880s, there was a distinguished mathematician, Leopold Kronecker, who philosophically disputed the existence of irrational numbers. At each step the numbers went through a period of experimentation and trial and were finally accepted only when some mathematician was able to develop a consistent system which contained both the old and the new numbers.

The same pattern holds for the infinitesimals. Even after their existence had been denied, they were in constant use as formal symbols. After 300 years of such usage, their existence finally became established when Robinson developed a system containing both infinitesimals and the real numbers. Robinson's system is now called the *hyperreal number system*, and using it he was able to completely justify the Leibniz approach to the calculus.

Often the new numbers which mathematicians invent shed light on the old numbers. For example, complex numbers were very useful in understanding real numbers. We will find that this is precisely the case with the hyperreals. They will be used exclusively for proving the theorems of calculus, theorems about real numbers. The power and beauty of this method, compared to the theory of limits, is sometimes astonishing. As Leibniz knew, the method of infinitesimals is the easy, natural way to attack these problems, while the theory of limits represents the lengths to which mathematicians were willing to go to avoid them.

Before actually launching ourselves into the details of the calculus, let us conclude this introductory section by considering an example of the second sort of physical problem solvable by calculus, namely the problem of finding the area of a plane figure. We might as

but of *discovering* them. This method is precisely the method of infinitesimals!

Although this letter was unknown to seventeenth-century mathematicians, they suspected that the Greeks were hiding something. To quote Leibniz: "these theorems they completed with *reductio ad absurdum* proofs by which they at the same time provided rigorous demonstration and also concealed their methods."

It is not likely that Archimedes was deliberately misleading; rather he felt the "method of exhaustion" was the only legitimate means for proving the theorem. Less honorable motives, however, may be assigned to later mathematicians:

I have omitted a number of things that might have made it [the geometry] clearer, but I did this intentionally, and would not have it otherwise. The only suggestions that have been made concerning changes in it are in regard to rendering it clearer to readers, but most of these are so malicious that I am completely disgusted with them. René Descartes (1596–1650)

And it is said Newton himself had difficulty with Analytic Geometry!

well use our function from before, so let A denote the region bounded by the curve $y = x^2$ and the lines $x = 1$, $x = 2$, and $y = 0$. We will find the area of A:

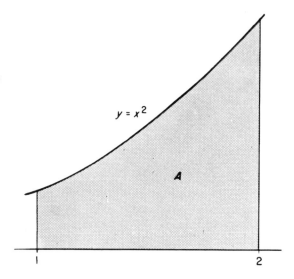

Our approach to the problem is very simple: Since we know how to find areas of rectangles, we will simply approximate the area of A by placing a great many thin rectangles over the region and adding up these areas. How thin should these rectangles be? Infinitesimally thin! We proceed as follows:

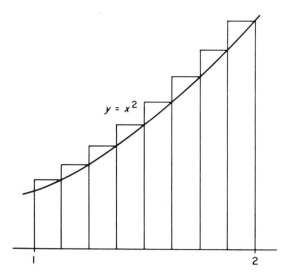

As in our previous example, we must leave out many details from this argument simply because we have not yet formally defined the infinitesimals or described their properties.

If we divide the area up into rectangles of thickness h, *where h is a real number*, not an infinitesimal, we

The ith rectangle has base h and height $(1 + ih)^2$:

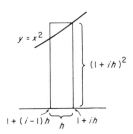

so the area is $h + 2ih^2 + i^2h^3$. Thus the sum of the areas of all the rectangles is

$$\sum_{i=1}^{n} (h + 2ih^2 + i^2h^3),$$

where $n = 1/h$ is the number of rectangles. Using the summation formulas derived later, it can be routinely calculated that this sum is

$$\frac{7}{3} + \frac{3h}{2} + \frac{h^2}{6}.$$

We will do such calculations in complete detail in Chapter 6 (see exer. 4, p. 58).

The calculus was the subject of lighthearted attacks as well as serious ones. An example is the satirical play by John Gay (author of the *Beggar's Opera*) called *Three Hours After Marriage*. The main character was a mathematician resembling Newton who spent much of his time at alchemy. (Actually, Newton *did* spend much of his time at alchemy, trying to turn base metals into gold.)

can use formulas from high school mathematics to show that the sum of the areas of the rectangles is

$$\frac{7}{3} + \frac{3h}{2} + \frac{h^2}{6}.$$

Does this formula still hold if h is ◎, an infinitesimal? As it happens, it does, and so to find out the desired area, we note that if ◎ is an infinitesimal, so are 3◎/2 and ◎²/6. Thus

$$\frac{7}{3} + \frac{3◎}{2} + \frac{◎^2}{6}$$

is infinitely close to 7/3. But since the area of A must be a real number, we argue, as we did in the tangent example, that the area must be 7/3.

There is one crucial point in our area calculation that bears repeating. We derived the formula for the approximation

$$\frac{7}{3} + \frac{3h}{2} + \frac{h^2}{6}$$

for h *an actual real number*, but we then assumed the formula was true even if h were ◎, an infinitesimal. With this the proof was simple and easy. In the next chapter we will construct what we call the hyperreal number system by adding new numbers, infinitesimals, to the reals, and *the most important part of our work will be to guarantee that formulas that work for reals also work for hyperreals, including infinitesimals.*

To accomplish this, we will have to have a very good idea of what we mean by "formula," and this is where the techniques of mathematical logic will come in. Earlier mathematicians who attempted to build the hyperreals were defeated by this idea. There are many, many formulas, in fact, infinitely many, and they seem to be in complete disorder. It was only with the concept of a *mathematical language* that Robinson was able to bring order out of chaos. With this one key idea and its twin concept, mathematical structure, it will be relatively easy for us to construct the hyperreals. We will do it slowly and carefully, leaving no loose ends, and when we are done we can attack the problems of the calculus with directness and ease.

Ironically, Leibniz, in addition to his development of the calculus and his contributions to law, religion, philosophy, and diplomacy, also proposed another calculus, a *calculus ratiocinator*. It would consist, he imagined, of "a general method in which all truths of the reason would be reduced to a kind of calculation." In this dream Leibniz anticipated by almost 200 years the birth of mathematical logic.

2

Language and Structure

I have so many ideas that may perhaps be of some use in time if others more penetrating than I go deeply into them someday and join the beauty of their minds to the labor of mine.
Gottfried Wilhelm Leibniz (1646–1716)

The job before us is to build the hyperreal numbers. At first glance this does not seem so difficult; all we have to do is take the real numbers and add some infinitesimals. There is, however, a hitch. As mentioned in Chapter 1, we want something very special out of the hyperreals. We want every formula that is true in the reals to be true in the hyperreals. This will not be easy. There are infinitely many formulas, one more complicated than the next, and yet we must construct our system in only a few pages.

This is where mathematical logic comes in. Without it, mathematicians tried in vain for 300 years to construct the hyperreals. With it, Robinson was able to surmount the difficulties with astonishing ease.

Language

. . . the Symboles serve only to make men go faster about, as greater Winde to a Winde-mill.
Thomas Hobbes (1588–1679)

Hobbes's comment refers to Wallis's *Arithmetica Infinitorum*, which he called a "scab of symbols." Wallis was a mathematician of considerable powers, while Hobbes was only a dilettante and a critic. Hobbes later believed he had squared the circle— a task proved impossible in the nineteenth century.

The key to our construction is to study first the language of the formulas. In this way we can organize the formulas and see how they arise. It sounds a little messy, but it isn't. When we actually build the hyperreals, it will be quite smooth and natural.

There are many different so-called mathematical systems, and each system has an appropriate language. The best way to understand them is to look at a number of examples.

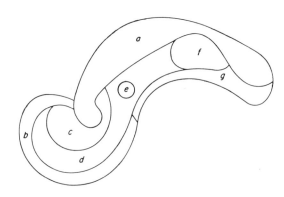

Example 1. Suppose we wish to study the diagram above and, in particular, the relationships between the different regions. The first requirement for our language is that it have symbols for the oddly shaped areas. Let's use the following symbols:

a

b

c

d

e

f

g

Leibniz was, among other things, the discoverer of the base 2 system for writing numbers. Leibniz was so impressed that all numbers could be written with only two symbols, 0 and 1, that he saw in this fact the presence of God. He later sent the idea to missionaries in China so that they could use it for purposes of conversion.

(Such symbols are commonly known as *constant symbols*.) Let's assume that we are primarily interested in the question of which areas are next to which. For this we use a *relation symbol* $N(\dots, ---)$ to mean "... is next to ---." For example, $N(c, b)$ says that c is next to b. Using these symbols we can make all manner of statements—*true*: $N(e, d)$, $N(g, b)$; and *false*: $N(c, g)$, $N(f, e)$. We will agree that no area is next to itself, so $N(c, c)$ is considered *false*.

To make longer and more intricate statements, we use the following *connectives*:

$\wedge$

$\vee$

$\rightarrow$

The symbol $\wedge$ means "and." For example, $N(c, d) \wedge N(f, a)$ is read "c is next to d *and* f is next to a." We can take any two statements P and Q and put them together to get $P \wedge Q$. This new statement is true if and only if *both P and Q are true*. For example,

$N(c, d) \wedge N(f, a)$ is true.
$N(b, f) \wedge N(d, e)$ is false.

This symbol was originally chosen because the latin word for "or," *vel*, begins with v.

There are many ways we could construct a language. We could use symbols different from ∧, ∨, and →; we could use different meanings for the symbols; etc. We have chosen this language because it is fairly easy to read and in standard use. As it happens, practically any reasonable language would work as well.

One example of a possible change in our language is this: Some people would suggest that "*P* or *Q*" should mean "either *P* or *Q*, but not both." This is a different kind of "or," and the symbol commonly used for it is "∨." We *could* use this instead of "∨," but it is less convenient, and we won't.

This definition of implication is difficult for some people to accept. The idea that a false statement "implies" any statement at all (if *P* is false, then *P* → *Q* is true no matter what *Q* is) is often challenged. At a party the great mathematician and philosopher Bertrand Russell tried to explain this point to a particularly obstinate individual who finally agreed to accept it if Russell could show that $0 = 1$ implied that Russell was the Pope. Russell reflected briefly and then argued: if $0 = 1$, then $1 = 2$. Since I and the Pope are two, I and the Pope are one. Q.E.D.

We could actually get by without all the symbols ∧, ∨, →, ~, but our sentences would be much harder to read. For example, instead of

$P \to Q$

we could say $\sim P \lor Q$. These two sentences mean the same thing.

∨ means "or." For example, $N(c, f) \lor N(a, b)$ is read "*either c is next to f or a is next to b*." Again, we can take any two statements *P* and *Q* and put them together to get $P \lor Q$, and this new statement is true if and only if *at least one of the two statements P or Q is true.* For example,

$N(b, f) \lor N(d, e)$ is true.
$N(c, d) \lor N(b, a)$ is true.
$N(b, f) \lor N(c, f)$ is false.

The symbol → means "implies." For example, $N(c, d) \to N(g, f)$ is read "*c is next to d implies g is next to f,*" or equivalently "*if c is next to d, then g* must be next to *f.*" Once again, we can take any two statements *P* and *Q* and put them together to get $P \to Q$. This statement is true if and only if *whenever P is true, so is Q.* Strictly speaking, the statement is only false when *P* is true and *Q* is false. For example,

$N(a, f) \to N(e, f)$ is false.
$N(a, f) \to N(d, f)$ is true.
$N(e, f) \to N(d, f)$ is true.
$N(e, f) \to N(b, e)$ is true.

Using parentheses ( ) the connectives can be used to make longer statements. For example,

$N(c, d) \quad \to (N(c, g) \lor N(f, a))$ (true).
$N(c, f) \quad \lor (N(b, g) \to N(b, f))$ (false).
$(N(b, a) \lor N(e, f)) \to (N(c, g) \land N(e, a))$ (false).

Finally, another symbol is useful. The symbol ~ means "not" or "it is not true that." The symbol ~ changes a statement from false to true or from true to false. For example,

$\sim N(d, f)$ is false.
$\sim N(g, c)$ is true.

EXERCISES
Are the following statements true or false?
1. $N(d, e) \land \sim N(b, f)$.
2. $N(c, e)$.
3. $\sim N(c, e)$.
4. $\sim \sim N(c, e)$.
5. $\sim \sim \sim N(c, e)$.
6. $\sim (N(g, c) \to N(g, f))$.
7. $\sim (\sim N(f, c) \land \sim N(g, b))$.
8. $(N(b, c) \lor \sim N(b, c)) \to (N(b, c) \land \sim N(b, c))$.

Similarly, instead of $P \wedge Q$, we could say

$\sim (\sim P \vee \sim Q)$.

Thus (with a great deal of difficulty) we could use just two symbols, $\sim$ and $\vee$, instead of four.

EXERCISES
1. Convince yourself that $P \rightarrow Q$ and $\sim P \vee Q$ mean the same thing.
2. Convince yourself that $P \wedge Q$ and $\sim (\sim P \vee \sim Q)$ mean the same thing.

For one interested in the "algebra" of logical connectives, it is possible to define a single connective that can replace all the others. Let "?" be defined by "$P ? Q$ is true if and only if *both P and Q are false*."
Using "?" we can replace $\sim$, $\vee$, $\wedge$, and $\rightarrow$. For example, $\sim P$ is true exactly when $P?P$ is true. Similarly, $P \vee Q$ can be replaced by

$(P ? Q) ? (P ? Q)$.

EXERCISES
1. Find a way to replace $\wedge$ by ?.
2. Find a way to replace $\rightarrow$ by ?.
3. Invent "¿."

The discovery of "?" was first made by an American, C. S. Peirce, in 1880, though he probably used a different symbol. This and much of Peirce's work remained unnoticed by mathematicians. Peirce explained this by saying "My damned brain has a kink in it that prevents me from thinking as other people do."
"?" was rediscovered by H. M. Sheffer in 1913.

Mathematicians are like Frenchmen: whatever you say to them they translate into their own language and forthwith it is something entirely different.
Johann Wolfgang von Goethe
(1749–1832)

9. $N(a, b) \rightarrow (N(b, c) \rightarrow (N(c, d) \rightarrow N(d, e)))$.
10. $\sim ((\sim N(g, a) \rightarrow N(e, f)) \vee \sim (\sim N(c, d) \wedge \sim (N(d, a) \wedge \sim N(f, b))))$.

For greater flexibility we can add *variables* x, y, and z to our language. These enable us to talk about objects without having to specify just which object we have in mind. For example, $N(x, a)$ is read "x is next to a" (which turns out to be true for some x and false for others), and $N(x, a) \vee N(x, d)$ is read "x is next to a or x is next to d," which is true for all x.

EXERCISES
For which areas x are the following statements true?
11. $N(x, e)$.
12. $\sim N(x, a) \wedge \sim N(x, c)$.
13. $\sim N(x, d)$.
14. $\sim N(x, a) \wedge N(x, g)$.
15. $\sim N(x, b) \wedge N(x, c)$.

Example 2. Let us study the genealogical table on page 17. We are interested this time in three relationships: the two-place relation "—is a parent of—," the two-place relation "—is married to—," and the one-place relation "—is a female."

Our language must include constants for all the people in the chart. Rather than use complete names, we will use the following abbreviations: OG for Owain Gwynedd, C for Christina, etc.

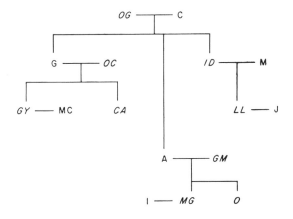

Thus our constants are OG, C, G, OC, ID, M, GY, MC, CA, LL, J, A, GM, I, MG, and O. In addition we will have variables x, y, and z, the same connectives and grammatical aids as before, $\wedge$, $\vee$, $\rightarrow$, $\sim$,), (, and three new relation symbols:

$M(x, y)$—"x is married to y"

$P(x, y)$—"x is a parent of y"

$F(x)$—"x is female"

Three Generations of Lords of Wales

All males trace their lineage to Merfyn Frych (the freckled).

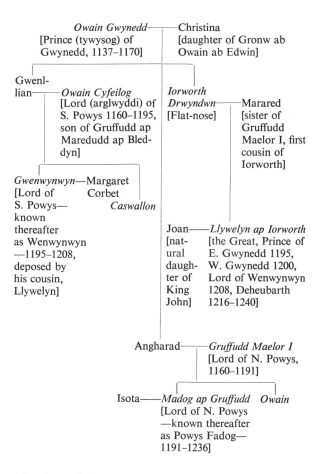

All males are in italics.

Source: *Handbook of British Chronology,* Sir F. Maurice Powicke and E. B. Fryde, ed. (Royal Historical Society, 1961). Note: The last independent Welsh prince was Llywelyn the Last, grandnephew of Llywelyn ap Iorworth.

To further enrich our language, we add the "$=$" sign and two *quantifiers*:

∀

∃

You can think of "$=$" as another 2-place relation, but we write it in the usual way. For example, $x = \text{GM}$ instead of $= (x, \text{GM})$, and $x \neq \text{GM}$ instead of $\sim =$

(x, GM). The quantifiers may also be familiar. $\forall$ means "for all," and it is used with a variable. For example,

$$\forall x \sim (M(x, \text{O}))$$

is read "for all x, x is not married to Owain," that is, Owain is unmarried, which is true. Another example:

$$\forall y(P(\text{G}, y) \to \sim F(y))$$

is read "for all y, if Gwenllian is a parent of y, then y is not female," which is also true. $\exists$ means "there exists," and it is also used with a variable. For example,

$$\exists x P(\text{ID}, x)$$

is read "there exists an x such that Iorworth Drwyndwn is a parent of x"—a true statement; and

$$\exists z(P(z, \text{J}) \wedge P(z, \text{O}))$$

is read "there exists z such that z is a parent of Joan and a parent of Owain"—a false statement.

One last remark: Is the statement

$$x = \text{J}$$

true or false? Neither, of course, since we don't know what x is. For this reason we give the following definition.

DEFINITION: A statement in a language is said to be a *sentence* if every variable appearing in it

$$\dots x \dots$$

is referred to by a quantifier

$$\forall x(\dots x \dots)$$

or

$$\exists x(\dots x \dots).$$

Thus, we could make "$x = \text{J}$" into a sentence by attaching a quantifier, getting either

$$\forall x(x = \text{J}) \qquad \text{(false)}$$

or

$$\exists x(x = \text{J}) \qquad \text{(true)}.$$

It should be clear that any sentence is always either true or false.

Here is another example of unnecessary extra symbols: $\forall x(\ldots)$ can be replaced by $\sim(\exists x \sim(\ldots))$. Both of these mean the same thing if you think about it.

EXERCISES
1. Convince yourself that $\forall x(\ldots)$ and $\sim \exists x \sim(\ldots)$ mean the same thing.
2. Convince yourself that $\exists x(\ldots)$ and $\sim \forall x \sim(\ldots)$ mean the same thing.

13′. x is *the* brother of y.

Define a new 2-place relation symbol $L(x, y)$ meaning "x loves y."

EXERCISES
1. Write in the language "Everybody loves somebody."
2. Write in the language "All mankind loves a lover"—Emerson (interpret this as you will).

EXERCISES

In each statement discover who x is.
1. $M(x, GY)$.
2. $P(M, x)$.
3. $P(C, x) \wedge \sim F(x)$.
4. $P(GM, x) \wedge \forall y \sim M(x, y)$.
5. $P(OC, x) \wedge \sim (x = GY)$.
6. $\exists y P(y, x) \wedge \exists z P(x, z) \wedge \sim F(x)$.
7. $F(x) \wedge \exists y[P(x, y) \wedge \forall z(P(x, z) \rightarrow z = y)]$.
8. $\sim F(x) \wedge \exists y \exists z[P(x, y) \wedge P(y, z)]$.
9. $F(x) \wedge \exists y[P(OC, y) \wedge P(x, y)]$.
10. $\exists y[P(y, x) \wedge P(y, LL)]$.
11. $\sim F(x) \wedge \exists y \exists z [P(x, y) \wedge P(x, z) \wedge [F(y) \rightarrow \sim F(z)] \wedge [F(z) \rightarrow \sim F(y)]]$.
12. $\exists y[F(y) \wedge M(y, x) \wedge P(y, x) \wedge \sim F(x)]$.

Allowing additional variables $x_1, x_2, x_3, \ldots$, write statements defining the following relationships:
13. x is a brother of y. For example, $(\sim F(x) \wedge \exists x_1 (P(x_1, x) \wedge P(x_1, y))) \wedge (x \neq y)$.
14. x is a grandfather of y.
15. x is a cousin of y.
16. x is a brother-in-law of y. (Careful—there are two different ways of being a brother-in-law.)
17. x is a niece of y. (This is also tricky.)
18. x is a stepmother of y.
19. x is a half brother of y.
20. x is a bastard.

Example 3. This diagram represents pictorially a function F. An arrow going from one point x to another point y means that the value of the function at x is y. For example, $F(a) = b$, $F(d) = c$, and $F(f) = f$. Our

One of the earliest studies of logic was Aristotle's, and it too involved the study of a special language. By comparison it was crude, allowing only statements of the form:

All_____are_____.
No_____are_____.
Some_____are_____.
Some_____are not_____.

From such sentences were formed arguments called syllogisms. Here is a typical syllogism:

All dogs are animals.
No fish are dogs.
All fish are animals.

The arguments were analyzed to discover whether they were valid or invalid. (The above is invalid.) Most of the work in logic for the next 2,000 years consisted of classifying and analyzing such arguments—which is less a tribute to Aristotle than a reflection on mathematics in the Middle Ages.

As late as the nineteenth century, mathematicians still looked for new ways to solve syllogisms. Robert Venn invented his Venn diagrams for just this purpose (you may have studied them in high school). Charles Dodgson (author of *Alice in Wonderland* under the name Lewis Carroll) also invented a method which he published under the title *The Game of Logic*.

The term "function" ("funktion" in German) was first coined by Leibniz.

Please note that the language described here is no more complicated than previous ones; it is simply bigger.

language will use constant symbols a, b, c, d, e, f, g, variables x, y, z, connectives $\wedge$, $\vee$, $\rightarrow$, $\sim$, quantifiers $\forall$, $\exists$, and a *function symbol F* standing for the function.

EXERCISES
State whether the sentence is true or false.
1. $\forall x \exists y (F(x) = y)$.
2. $\forall x \exists y (F(y) = x)$.
3. $\exists x \forall y (F(x) = y \rightarrow x = y)$.
4. $\forall x (F(x) = x \rightarrow (x = f \vee x = g))$.
5. $\forall x (\exists y \exists z (F(y) = x \wedge F(z) = x \wedge \sim (y = z)) \rightarrow (x = c \vee x = g))$.
6. $\exists x \exists y \exists z ((\sim (x = y) \wedge F(x) = y) \wedge (F(y) = z \wedge F(z) = x))$.
7. $\forall x ((F(x) = F(F(x))) \rightarrow F(F(x)) = F(F(F(x))))$.

In each of our examples we have had the following categories:
1. constants
2. variables
3. grammatical symbols and connectives: $\vee$, $\wedge$, $\rightarrow$, $($, $)$
4. quantifiers: $\forall$, $\exists$
5. relation symbols (including "$=$")
6. function symbols.

In the future our languages will be constructed from *only* these categories. All will have the first four categories and will generally have additional relation and function symbols. All our languages will have at least the relation symbol "$=$."

Our next example is the most important one to us. It is a language to describe the real number system, and it is this language upon which our definition of the hyperreals will ultimately depend.

The formal construction of mathematical languages was accomplished as recently as the last century. Leibniz made some attempts that are, in retrospect, astonishingly modern, but they were ignored. Leibniz's dream was that all thought could be written in such a language: ". . . *when controversies arise, there will be no more necessity of disputation between two philosophers than between two accountants. Nothing will be needed but that they should take pen in hand, sit down with their counting tables and (having summoned a friend, if they like) say to one another:* Let us calculate."

Modern logic began with George Boole (1815–1864) who first experimented with the languages we are using here. Boole called it an algebra, because ∨ reminded him of addition, and ∧ reminded him of multiplication. For example, $P \wedge Q$ and $Q \wedge P$ mean the same — ∧ is "commutative." Similarly,

$$P \wedge (Q \wedge R) \text{ and } (P \wedge Q) \wedge R$$

mean the same — ∧ is "associative." Similarly, ∨ is both commutative and associative.

Finally, the distributive law which states

$$a \cdot (b + c) = (a \cdot b) + (a \cdot c)$$

is also true in logic, that is, $P \wedge (Q \vee R)$ means the same as $(P \wedge Q) \vee (P \wedge R)$.

It is a good exercise to convince yourself of these facts.

To get the reader started on these exercises, the answers to the first two are:

1. $\forall x_1 (x_1 = x_1)$.
2. $\forall a \, \forall b \, (a = b \rightarrow b = a)$.

There are properties of the real numbers which cannot be expressed in L. Chief among these is the *axiom of completeness*.

Example 4.

DEFINITION. *The language L for the real numbers* will consist of:

constants. A constant symbol "r" for each real number r.

variables. x_1, x_2, x_3, ... and a, b, c, ...

grammatical symbols, connectives, quantifiers.

functions. We certainly want our language to have function symbols for addition, subtraction, multiplication, and division. For these we will use the symbols $+$, $-$, $\times$, $\div$. In addition to these we will need many other function symbols. For example, we will need ones for sine, cosine, logarithms, etc. To play it safe, for every function f on the real numbers, we include a function symbol f in the language L.

relations. We include the two-place relations $=$ and $<$, of course, but there are others. We will want to express, for example, "x is an integer," and for this we introduce a 1-place relation symbol $I(x)$ meaning "x is an integer." We might also need one saying "x is a rational number," and there are many other relations we might or might not need. To be prepared, for *every* relation R on the real numbers, we throw an associated relation symbol R into L.

EXERCISES

The following are most of the axioms for the real numbers, properties that we have seen many times before and accept without proof. Write them in the language L.

1. Every number is equal to itself.
2. If a is equal to b, then b is equal to a.
3. If a and b are both equal to c, then they are equal to each other.
4. Addition is commutative.
5. Addition is associative.
6. Zero is the additive identity.
7. Every number has an additive inverse.
8. Multiplication is commutative.

DEFINITION. *r* is an *upper bound* for a set of reals *B* iff *r* is greater than or equal to all elements of *B*. *r* is the *least upper bound* of *B* if *r* is an upper bound, and no other upper bound for *B* is less than *r*.

THE AXIOM OF COMPLETENESS. For all nonempty sets of numbers *B*, if *B* has an upper bound, then *B* has a least upper bound.

This axiom cannot be written in *L* because the phrase "for all nonempty sets of numbers" cannot be translated. We can quantify *numbers* in *L*, but not *sets of numbers*.

9. Multiplication is associative.

10. 1 is the multiplicative identity.

11. Every number not equal to 0 has a multiplicative inverse.

12. Multiplication is distributive over addition.

13. The sum of two positive numbers is positive.

14. The product of two positive numbers is positive.

15. The additive inverse of a positive number is not positive.

16. Every number not equal to 0 is either positive or negative.

Structure

The essence of mathematics lies in its freedom.
Georg Cantor (1845–1918)

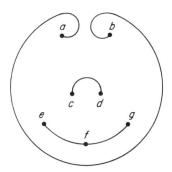

Cantor was one of the greatest mathematical explorers of the infinite, having the courage and foresight to go where all other mathematicians feared to tred. Perhaps his most startling discovery was that there are more irrational numbers than rational numbers.

Example 5. Let us take another example of a language. We will describe a party of seven people represented by the constants *a*, *b*, *c*, *d*, *e*, *f*, and *g*. We will have one relation symbol $N(x, y)$, which means "*x* shook hands with *y*." In the above diagram persons who shook hands are identified by a line drawn between them.

Looking back to page 14, we see that this language is *exactly* the same as the one in example 1. The identical language is being used in two different contexts. Consider the sentence

$N(e, f)$.

Is it true? Well, it is true about example 5 but false about example 1. This shows us that the truth of a statement depends on the context. In mathematics this idea of a "context" is known as structure. So far in this chapter, we have had examples of four different languages and five different structures (or contexts).

The rigorous formalization of truth in a structure is due to Alfred Tarski (1935).

DEFINITION. A *structure* $\langle S, R, F \rangle$ *appropriate to a given language* $\mathscr{L}$ consists of three things: S, a set of elements; R, a set of relations on S; and F, a set of functions on S; such that

1. Each constant of $\mathscr{L}$ corresponds to some element of S;
2. Each relation symbol of $\mathscr{L}$ corresponds to some relation on S in R; and
3. Each function symbol of $\mathscr{L}$ corresponds to some function on S in F.

In example 5 the set S was the group of people, there were no functions, and the relation was that in which people shook hands with each other.

Example 6. If the same seven people got together the next night and had a slightly different pattern of shaking hands,

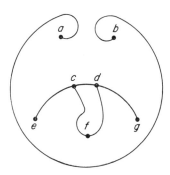

Essentially, structures are simply the things that mathematicians study. Easily, the integers form a structure. The real numbers form a structure. With minor effort, Euclidean geometry can be formed as a structure, as can the game of chess. In fact, our languages themselves can be formed into structures.

we would have a different structure because, although the set S is the same, we would have a different relation.

We generally give the structure with sets S, R, and F the name: $\langle S, R, F \rangle$. S may have elements that do not correspond to constants, but every constant must correspond to some element of S.

Example 7. Another structure appropriate to the language of examples 1 and 5 is the alphabet

$S = \{a, b, c, \ldots, x, y, z\}$
$R = \{\text{the alphabetical ordering}\}$
$F = \{\ \ \}.$

In this case $N(e, f)$ is true because e comes before f in the alphabet.

Conversely, although all constants must correspond to elements of S, they need not correspond to different elements.

The first coherent attempts to invent a comprehensive *mathematical* language were made independently by Gottlob Frege in 1879 and Giuseppe Peano in 1889. Later, Bertrand Russell and Alfred North Whitehead invented a language that was able to express virtually all known mathematical thought.

The logic we are working with here is called *2-valued logic*, because every sentence is either *true* or *false*. It is possible to invent more complex logics where sentences are either true, false, or partly true. We certainly don't need that here. Modern 3-valued logic was invented by Łukasiewicz in 1921. Oddly enough, it was discovered in 1936 that an English mathematician had also invented it—William of Occam (1270–1349)!

Example 8. Here is another structure appropriate to the same language:

$$S = \{17\}$$
$$R = \{\text{the relation that nothing is related to itself}\}$$
$$F = \{ \quad \}.$$

(That is, $N(17, 17)$ is false.) In this case *all* the constants, a, b, c, d, e, f, and g, correspond to 17, the only element of S, and $N(e, f)$ is false since both e and f represent 17. All of the structures in examples 1, 5, 6, 7, and 8 are appropriate to the same language.

In example 2, S is a set of 16 people, F is the empty set, and R contains three relations, the parent-child relation, the female relation, and the marriage relation. In example 3, S is a set of 7 letters, F contains the function described by the arrows, and R is empty. In example 4, S is the set of all real numbers, F is the set of all functions on the reals, and R is the set of all relations on the reals.

3

The Hyperreal Numbers

One cannot escape the feeling that these mathematical formulae have an independent existence and an intelligence of their own, that they are wiser than we are, wiser even than their discoverers, that we get more out of them than was originally put into them.

Heinrich Hertz (1857–1894)

We are now in a position to state precisely what we mean by a "hyperreal number system."

DEFINITION. A structure S is a hyperreal number system if it has the following three properties:

1. *S contains the real number system.* By this we mean not only that all real numbers are in S, but also that every function and relation defined on reals is also defined on numbers in S.

It's nothing new for one number system (or structure) to contain another. For example, the real number system contains the rational number system.

Concerning functions, we must have, for example, $\sin p$ defined for all hyperreal numbers p in S.

2. *S contains an infinitesimal.* That is, there is a number $\odot$ in S such that $\odot > 0$ and yet $\odot < r$ for every positive real number.

3. *The same sentences of L are true in both S and $\mathbf{R}$.* If B is any sentence of L, then B is true in S if and only if B is true in $\mathbf{R}$.

This is actually quite a special property. With the reals and the rationals, the sentence

$\exists\, x(x^2 = 2)$

is true in the reals but not in the rationals.

Warning: The language L is rather limited and is used solely for the purpose described in (3). It is neither sufficient nor appropriate for more general use. You will notice that this book is not written in L.

Notice that we defined what we mean by *a* hyperreal number system rather than *the* hyperreal number system. This is simply because there are many different hyperreal number systems. What is remarkable is that for the purpose of doing calculus any one of them is as good as any other.

In this chapter we shall construct a particular hyperreal number system denoted by $\mathbb{H}\mathbf{R}$. (Henceforth, we'll call this particular system "the hyperreals.") An understanding of the construction itself is not necessary for pursuing hyperreal calculus—one can fully understand the subsequent chapters of this book armed solely with a good comprehension of the above definition of a hyperreal number system.

At this point the reader may move immediately to Chapter 4, or continue with this chapter if he wishes a

It is important to note that although our language L will be strong enough to aid in the calculus, it will be too weak to distinguish between the reals and the hyperreals. For example, it is impossible to discuss infinitesimals in L.

EXERCISES

1. Let L_1 be the language with constants $1, 2, 3, 4, \ldots$ and the function $+$ (and all the usual variables, connectives, etc.). Write a sentence in L_1 that is true in $N = \{1, 2, 3, 4, \ldots\}$ but not true in $M = \{0, 1, 2, 3, 4, \ldots\}$. (*Hint*: Try to say N has no identity.)

2. Let L_2 be the same as L_1 but with the constant 0 added. Write a sentence in L_2 that is true in M but not in

$$I = \{\ldots, -2, -1, 0, 1, 2, 3, \ldots\}.$$

3. Write a sentence in L_2 that is true in Q = the rational numbers but not true in I.

4. Can you write a sentence in L_2 that is true in Q but not in $\mathbf{R}$?

5. Write a sentence in L_1 true in I but not in N. (not easy)

6. Write a sentence in L_1 true in I but not in M. (even harder)

7. Let L_3 be the language with no constants, no functions, and the relation $<$ (plus the usual stuff). Write a sentence in L_3 true in N but not in I.

8. Write a sentence in L_3 true in Q but not in I.

9. Can you write a sentence in L_3 true in M but not in N?

10. Can you write a sentence in L_3 true in Q but not in $\mathbf{R}$? (Don't spend more than two hours on this.)

This problem occurs in any base representation of real numbers. For

more detailed presentation of the hyperreal number system.

There is a loose analogy between our construction of the hyperreal numbers and the decimal construction of the reals. To motivate our construction let us explore this.

To get the reals we first take the integers

$$\ldots, -3, -2, -1, 0, 1, 2, \ldots,$$

and define *decimals*

$$587.718222222\ldots$$
$$437.006182034\ldots$$
$$-236.000000000\ldots$$
$$7.999999999\ldots.$$

We can think of a decimal as being a sequence of integers. The first number in the sequence can be any integer, such as 587, but all the rest (the digits after the decimal point) must be between 0 and 9. Thus the decimal number

$$587.7182222222\ldots$$

can be seen as the infinite sequence

$$587, 7, 1, 8, 2, 2, 2, 2, 2, \ldots.$$

By analogy it will turn out that the hyperreals will also be made from sequences—not sequences of integers, but sequences of real numbers!

The reals contain the integers, that is, every integer is also a decimal. Is 12 a decimal? Of course, but as a decimal we write it

$$12.000000000000\ldots.$$

In our sequence notation, 12 is

$$12, 0, 0, 0, 0, 0, 0, \ldots.$$

We will show that the real numbers are contained in the hyperreals in a similar way.

Finally, we have one peculiar problem with decimals. Two different looking decimals might actually be the same number. For example, .9999999 ... is the same as

example, in base 6,

$$1.0000 \ldots_6 = .5555 \ldots_6$$

and in base 2,

$$1.0000 \ldots_2 = .1111 \ldots_2.$$

This even occurs in negative number bases. For example, in base -2,

$$.11010101 \ldots_{-2} = .00101010 \ldots_{-2}$$

where $.11010101 \ldots_{-2}$ is

$$1 \cdot (-\tfrac{1}{2}) + 1 \cdot (\tfrac{1}{4}) + 0 \cdot (-\tfrac{1}{8}) + \cdots = -\tfrac{1}{6}$$

and $.00101010_{-2}$ is

$$0 \cdot (-\tfrac{1}{2}) + 0 \cdot (\tfrac{1}{4}) + 1 \cdot (-\tfrac{1}{8}) + \cdots = -\tfrac{1}{6}.$$

A more complete and rigorous discussion of bases and decimals will be given in Chapter 9.

Notice that we are referring to hyperreals both as "sequences" and as "functions." This is simply because sequences really are functions. The sequence

$$a_1, a_2, a_3, \ldots$$

corresponds to the function defined on the positive integers whose value at 1 is a_1, whose value at 2 is a_2, etc. Using function (rather than sequence) notation, the hyperreal q is represented by the function f given as follows:

$$f(n) = \begin{cases} 4 & \text{if } n = 1 \\ \tfrac{1}{2} & \text{if } n = 2 \\ 2 & \text{if } n = 3 \\ 1 & \text{if } n = 4 \\ 5 & \text{if } n \geq 5 \end{cases}$$

and p is represented by g:

$$g(n) = \begin{cases} 1 & \text{if } n = 1 \\ 7 & \text{if } n = 2 \\ 5 & \text{if } n \geq 3. \end{cases}$$

Thus two functions f and g will represent the same hyperreal if and only if

$$\{n \mid f(n) = g(n)\}$$

is quasi-big.

In the case of the hyperreals p and q mentioned earlier (represented by f and g), the set on which p and q agree is

$$\{n \mid f(n) = g(n)\} = \{5, 6, 7, 8, 9, 10, \ldots\}.$$

$1.0000000\ldots$, $3.9999999\ldots$ is the same as $4.0000000\ldots$, and $-5.3499999\ldots$ is the same as $-5.350000\ldots$.

This same problem will happen with the hyperreals but in a somewhat different form.

As we previously stated, the hyperreals will be sequences of real numbers. Here are a few examples:

a. $2, 1, \tfrac{1}{2}, \tfrac{1}{4}, \tfrac{1}{8}, \ldots$

b. $1, 2, 3, 4, 5, \ldots$

c. $2, 2, 2, 2, 2, \ldots$

d. $437, -19, 437, -19, 437, \ldots$.

We also said that many different sequences may represent the same hyperreal.

It will turn out, for example, that the sequences

$$p: 1, 7, 5, 5, 5, 5, 5, 5, 5, 5, \underbrace{\ldots}_{\text{all 5s}}$$

and

$$q: 4, \tfrac{1}{2}, 2, 1, 5, 5, 5, 5, 5, 5, \underbrace{\ldots}_{\text{all 5s}}$$

will represent the same hyperreal. The reason for this is that p and q are the same at almost every place, that is, they agree on a very large set.

To make this precise, we must settle in our minds what we mean by "a very large set." In the example above the two sequences agree at all but the first four places—this is certainly on a large set. In general we could adopt the rule that two sequences represent the same hyperreal if they agree on a so-called big set, that is, a set of natural numbers so large that it includes all natural numbers with the possible exception of finitely many. But this is not quite enough. What we must do is define what we call *quasi-big* sets. With these our job will be complete, for we will then say that "two sequences represent the same hyperreal iff the set of places where the two are the same is quasi-big."

What are the quasi-big sets? The rigorous definition is somewhat complex, and we have left it for Appendix A. You may read this now if you wish, but it isn't necessary in order to understand what follows. Indeed, in order to grasp the rest of this chapter, one need only keep in mind the following four properties of quasi-big sets:

1. No finite set is quasi-big.

2. If A and B are quasi-big, then so is $A \cap B$.

Actually, property (3) is implied by the others. If A is quasi-big and $A \subseteq B$, then B must be quasi-big, for suppose it isn't: then by (4), B^c, the complement of B, is quasi-big. Next, by (2), $A \cap B^c$ is quasi-big, but $A \cap B^c$ is empty, hence finite, so by (1) it cannot be quasi-big, a contradiction. In pictures:

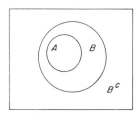

Since the set $A = \{5, 6, 7, 8, 9, ...\}$ consists of all natural numbers except a finite set, A is quasi-big and so p and q are equal hyperreals.

Recall that by definition, a structure consists of three things, a set of elements (in this case the hyperreal numbers), a set of functions, and a set of relations.

For example, suppose that j is

$1, \frac{1}{2}, \frac{1}{3}, \frac{1}{4}, \frac{1}{5}, ...$

and $f(x) = 2x + 1$ is our function. Then $f(j)$ is

$3, 2, 1\frac{2}{3}, 1\frac{1}{2}, 1\frac{2}{5}, ...$

For example, if $g(x, y)$ is the 2-place function $x^2 - y$, and

j: 2, 3, 4, 2, 3, 4, ...

k: $-2, -1, 0, 1, 2, 3, ...$

are hyperreals, then $g(j, k)$ is

3. If A is quasi-big and $A \subseteq B$, then B is quasi-big.
4. If A is any set, then either A or its complement is quasi-big.

Note that if A consists of all natural numbers except for finitely many, then by (4) and (1) A must be quasi-big.

We are now ready to get down to business. The hyperreal numbers themselves are just sequences of reals subject to the condition that *two sequences*

$i(0), i(1), i(2), ...$

and

$k(0), k(1), k(2), ...$

represent the same hyperreal iff

$\{n \mid i(n) = k(n)\}$

is quasi-big.

The hyperreal numbers constitute the domain of the structure we desire, but in and of themselves they say nothing about how various functions of hyperreals and relations of hyperreals operate. This now becomes our focus.

Suppose that f is a function on the *reals* in our language L. How do we define f on the *hyperreals*? Suppose that j is a hyperreal, What is $f(j)$?

Well, j is the sequence $j(1), j(2), j(3), j(4), ...$, so we define $f(j)$ to be the new sequence

$f(j(1)), f(j(2)), f(j(3)), f(j(4)), ...$

Here is a possible problem: suppose that j and k represent the same hyperreal. Then $f(j)$ and $f(k)$ should represent the same hyperreal. Do they? Certainly, for let

$A = \{n \mid j(n) = k(n)\}$

and

$B = \{n \mid f(j(n)) = f(k(n))\}.$

We know A is quasi-big and we can easily check that

$A \subseteq B.$

Thus B must be quasi-big and so $f(j)$ and $f(k)$ represent the same number. In this case f was a 1-place function, but 2-place, 3-place, and other functions work in the same way.

We now know how functions are defined. What about relations? Suppose $R(\ )$ is a relation and j is a

6, 10, 16, 3, 7, 13, ...

Similarly, to add j and k, we simply add them term by term:

$j + k$: 0, 2, 4, 3, 5, 7, ...

EXERCISES
1. Prove that addition is commutative in HR.
2. Prove that every hyperreal number has an additive inverse. (*Hint:* These are very easy.)
3. Let $j(1), j(2), j(3), \ldots$ be any hyperreal, and let $f(x)$ be the function $f(x) = \sin^2 x + \cos^2 x$. Show that $f(j) = 1$ (this proves that the rule $\sin^2\theta + \cos^2\theta = 1$ is true in HR).

For example, suppose $R(\ )$ is the relation "is greater than $2\frac{1}{2}$," and j is the hyperreal

3, 4, 3, 4, 3, 4, ...;

then $R(j)$ is true, since $R(j(n))$ is true for *all n*, so $\{n \mid R(j(n))$ is true$\}$ is quasi-big.

An important example of a 2-place relation is $<$. When is $j < k$? The answer is that $j < k$ whenever $\{n \mid j(n) < k(n)\}$ is quasi-big.

It is not hard to see that this way of writing real numbers poses no problems. For example, if r is $r, r, r, \ldots$ and s is $s, s, s, \ldots$, then $r + s$ is $r + s, r + s, r + s, \ldots$, and rs is $rs, rs, rs, \ldots$, and so on.

hyperreal. How do we decide whether $R(j)$ is true or false? Well, we simply define $R(j)$ to be true if and only if $\{n \mid R(j(n))$ is true$\}$ is quasi-big.

Here is another possible problem: Suppose j and k represent the same hyperreal. $R(j)$ should be true if and only if $R(k)$ is true. Is this right? Certainly, for let

$A = \{n \mid R(j(n))$ is true$\}$,

$B = \{n \mid R(k(n))$ is true$\}$,

and

$C = \{n \mid j(n) = k(n)\}$.

Suppose $R(j)$ is true. Then A is quasi-big. We know C is quasi-big, so $A \cap C$ must also be, and since $A \cap C \subseteq B$, B must also be quasi-big, that is, $R(k)$ is true.

Again, this is only a 1-place relation, but 2-place, 3-place, etc., relations work just the same. We have thus completely defined our structure for the hyperreal numbers.

Now that we have the hyperreals, we must show that they have the three properties that we want, namely:
1. They contain the reals.
2. They contain an infinitesimal ⊚.
3. Every sentence of L which is true in HR is true in **R**, and vice versa.

EXERCISES
Prove that the following laws are true in **R**:
1. If $j < k$, then $j \neq k$.
2. If $j > 0$ and $k > 0$, then $j + k > 0$.
(*Hint:* if
$A = \{n \mid j(n) > 0\}$
and
$B = \{n \mid k(n) > 0\}$
are quasi-big, then $A \cap B$ is quasi-big.)
3. For all $i, j,$ and k, if $i < j$ and $j < k$, then $i < k$.

Property 1 is quite simple. For just as the integers were contained in the reals (-17 becomes -17.000000 ...), the reals are contained in the hyperreals, any real r being represented as the hyperreal $r, r, r, r, r, \ldots$. For example, 3 is 3, 3, 3, 3, ..., $2\frac{1}{2}$ is $2\frac{1}{2}, 2\frac{1}{2}, 2\frac{1}{2}, \ldots$, and π is $\pi, \pi, \pi, \ldots$.

Also, if r is less than s, then is r, r, r, ... less than s, s, s, ...? Let i be the function representing r ($i(n) = r$ for all n) and j be the function representing s. Then by definition of relations on HR, $i < j$ if and only if $\{n \mid i(n) < j(n)\}$ is quasi-big. But $\{n \mid i(n) < j(n)\}$ is the entire set of natural numbers (as $i(n) = r < s = j(n)$ for all n), and the set of natural numbers is certainly quasi-big.

For example, is $\bigcirc\!\!\!\!\!\bigcirc < 1/1,000,000$? Let j represent $1/1,000,000$. Then

$$
\begin{aligned}
A &= \{n \mid \bigcirc\!\!\!\!\!\bigcirc \, (n) < j(n)\} \\
&= \{n \mid 1/n < 1/1,000,000\} \\
&= \{1,000,001 \quad 1,000,002 \\
&\quad 1,000,003 \,...\}
\end{aligned}
$$

is quasi-big since it contains all but finitely many natural numbers.

Notice that HR also contains infinite numbers: let j be the hyperreal 1, 2, 3, 4, 5, 6, ...

EXERCISES
1. Prove that j is greater than every real number.
2. Prove that j is actually a "hyperinteger," that is, in HR the sentence "j is an integer" is true.

We have already proved this for some sentences (in previous exercises). Here are some more. Prove that the following sentences of L are true in HR:
1. $\forall x \, (x^2 \geq 0)$.
2. $\forall x \, (x \neq 0 \rightarrow \exists y \, (x \cdot y = 1))$. (*Hint*: For any hyperreal j, let k be the hyperreal

$$
k(n) = \begin{cases} \dfrac{1}{j(n)} & \text{if } j(n) \neq 0 \\ 17 & \text{if } j(n) = 0. \end{cases}
$$

Then show that if $j \neq 0$, $j \cdot k = 1$.)
3. $\forall x \, \forall y \, (x \cdot y = 0 \rightarrow (x = 0 \lor y = 0))$. (*Hint*: For hyperreals j and k, draw a Venn diagram showing the sets

$A = \{n \mid j(n) = 0\}$,
$B = \{n \mid k(n) = 0\}$,
and
$C = \{n \mid j(n) \cdot k(n) = 0\}$.)

We would next like to check that **R** contains infinitesimals. Here is a possibility:

$1, \frac{1}{2}, \frac{1}{3}, \frac{1}{4}, \frac{1}{5}, \frac{1}{6}, \frac{1}{7}, \dots$.

Let's call this function $\bigcirc\!\!\!\!\!\bigcirc$. Is $\bigcirc\!\!\!\!\!\bigcirc > 0$? Yes, because each element in the sequence is > 0, that is, if i represents $0 = 0, 0, 0, 0, 0, \dots$ then $\{n \mid \bigcirc\!\!\!\!\!\bigcirc(n) > i(n)\}$ is the set of *all* natural numbers, hence quasi-big.

Is $\bigcirc\!\!\!\!\!\bigcirc < r$ for every positive real number r? Yes, for suppose r is any positive real. Then if j represents $r = r, r, r, r, \dots$, let's look at

$A = \{n \mid \bigcirc\!\!\!\!\!\bigcirc(n) < j(n)\}$.

Is A quasi-big? Well, no matter how small r is, we can always find an integer k such that

$$
\frac{1}{k} < r,
$$

and of course this also means that

$$
\frac{1}{k+1} < r, \quad \frac{1}{k+2} < r, \dots.
$$

Thus, $\{k, k+1, k+2, \dots\} \subseteq \{n \mid \bigcirc\!\!\!\!\!\bigcirc(n) < j(n)\}$ and hence $\{n \mid \bigcirc\!\!\!\!\!\bigcirc(n) < j(n)\}$ is quasi-big. Thus, $\bigcirc\!\!\!\!\!\bigcirc < r$ and since r was arbitrary, $\bigcirc\!\!\!\!\!\bigcirc$ must be an infinitesimal.

Our third and last problem is more difficult. We must show that every sentence in L true in HR is also true in the real numbers and vice versa. To do this we examine our language L. L consists of

1. a constant symbol r for every real number
2. a function symbol f for every real function
3. a relation symbol R for every real relation.

In order to talk about hyperreals, we must expand L a little bit to include

4. a constant symbol for every *hyperreal* number.

We will call the new language L^*.

Notation: If G is a formula of L^*, and G uses the new symbols $j_1, j_2, j_3, \dots, j_k$, then we will sometimes write

$G(j_1, j_2, j_3, \dots, j_k)$,

which only means that $j_1, j_2, j_3, \dots, j_k$ are the only new constants in the formula G.

Notice that since $j_1, j_2, j_3, \dots, j_k$ are really sequences, $j_1(n), j_2(n), j_3(n), \dots, j_k(n)$ are just real numbers, so that the formula $G(j_1(n), j_2(n), \dots, j_k(n))$ (which means: put $j_1(n)$ instead of j_1, $j_2(n)$ instead of j_2, ..., etc.) doesn't

4. Any of the statements in the exercises on pp. 21–22.

This theorem is due to Łos.

Some persons have contended that mathematics ought to be taught by making the illustrations obvious to the senses. Nothing can be more absurd or injurious: it ought to be our never-ceasing effort to make people think, not feel.[1]
Samuel Taylor Coleridge (1772–1834)

use any new symbols of L^*. Thus it is actually written in L.

DEFINITION. If $G = G(j_1, j_2, ..., j_k)$ is a formula of L^*, then G_n is the formula of L

$$G(j_1(n), j_2(n), ..., j_k(n)).$$

We can now state the theorem that will finish the job.

THEOREM 3.1. If G is any formula of L^*, then G is true in HR if and only if $\{n \mid G_n$ is true in $\mathbf{R}\}$ is quasi-big.

This theorem is a special case of a very important theorem of logic called the fundamental theorem of ultraproducts. The proof is not difficult, but a little long, so it is placed in Appendix B.

Let us see why this theorem proves what we want: Suppose G is a sentence not only of L^*, *but also of L.* This means that G doesn't have any hyperreal numbers appearing in it, and so we must have $G_n = G$ for all n. Thus, if G is in L, $\{n \mid G_n$ is true in $\mathbf{R}\}$ is either the empty set (if G is false in $\mathbf{R}$) or the set of all the natural numbers (if G is true in $\mathbf{R}$). The empty set is not quasi-big, and the set of all natural numbers is, and so, immediately by our theorem, G is true in HR iff G is true in $\mathbf{R}$. This completes our construction of the hyperreals.

---

[1]It should be noted that the authors do not agree with this.

4

*So Nat'ralists observe, a Flea Hath
Smaller Fleas that on him prey,...*
Jonathan Swift (1667–1745)

The Hyperreal Line

In this chapter our goal is to acquire a good intuitive feeling for the hyperreal numbers. As of now all we know is that the same sentences of L are true in $\mathbb{HR}$ and $\mathbb{R}$—this hardly entails a good practical grasp of the numbers themselves.

What do the hyperreals look like? The usual way to picture the real numbers

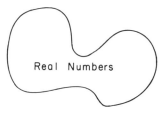

is as a set of points sitting on an infinitely long line:

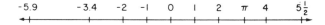

Such a picture is extremely natural and helps us to develop a good "intuitive feeling" about the reals. Thus it would be particularly nice if we knew that we could picture the hyperreals

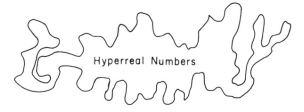

as points sitting on a line. Fortunately we can.

What is it about *real* numbers that allows us to view them as points on a line? A moment's thought should tell us that there are two properties. The first is that given any two real numbers a and b, $a \neq b$, either a is less than b

*The advancement and perfection of
mathematics are intimately connected
with the prosperity of the State.*
Napoleon (1769–1821)

or else b is less than a

but not both. This tells us that given any two reals we can decide which one "sits to the right of" the other on the line.

The other property says that given any three reals a, b, and c, if a is less than b

These two properties define what is called a *linear ordering* because the objects thus ordered can be put in a line. For instance, we can order the set of all kings of England by: One king $<$ another king if and only if the first king ruled earlier. This is clearly a linear ordering. An example of an ordering that is not linear: Let us order the set of subsets of the set $\{a, b, c\}$ by: One set $<$ another set if and only if the first set is contained in the second. For example,

and b is less than c,

then a is less than c.

$\{a\} < \{a, c\}$.

This is not a linear ordering because the first property of linear orderings is not satisfied. For example, we have

$\{a\} \not< \{b\}$

and

$\{b\} \not< \{a\}$.

Thus we can't picture this subset ordering as a line. In fact, it would look like this:

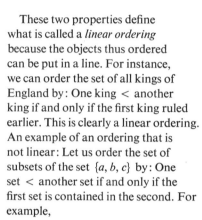

EXERCISE
Invent a set and an ordering on that set which satisfies the first property of linear orderings but not the second.

Thus if b sits to the right of a and c sits to the right of b, then c sits to the right of a.

We would like the hyperreal numbers to have these two properties. Do they? Certainly, since they can both be written as sentences in the original language of the reals, the first as

$\forall x_1\ \forall x_2(x_1 \neq x_2 \rightarrow (x_1 < x_2 \lor x_2 < x_1) \land \sim(x_1 < x_2 \land x_2 < x_1))$

and the second as

$\forall x_1\ \forall x_2\ \forall x_3\ ((x_1 < x_2 \land x_2 < x_3) \rightarrow x_1 < x_3)$.

Each of these sentences is true of the reals, and since ℝ was constructed so that every sentence of L true in the reals is true in ℝ, these two sentences must be true in ℝ. Thus the hyperreals have the two properties and can be pictured as points sitting on a line. Here is our view of ℝ at this point:

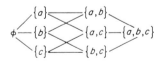

What numbers are in ℝ? As of now we know that ℝ contains the reals and ⊚, an infinitesimal. Thus our picture really looks like this:

Note that since ◎ is greater than 0 and yet less than every positive real, it must be a hyperreal number that is not a real number. There are two definitions appropriate to describe ◎:

DEFINITION. A hyperreal is said to be *nonstandard* if it is not real.

Note that every infinitesimal must be nonstandard.

DEFINITION. A hyperreal is an *infinitesimal* if it is not equal to 0, yet is smaller than every positive real and larger than every negative real.

Let us now see if there are any infinitesimals besides ◎ in HR. Any hyperreal ∗ that satisfies

$$0 < * < ◎$$

During the first two hundred years of the calculus, infinitesimals were not well understood. They were handled rather carelessly and often confused with the number 0. For this reason this period of mathematical history was once called "the golden age of nothing."

Early theories of infinitesimals considered such numbers to be "indivisible," that is, "atomic particles" of the real line. (It was even stated during the Middle Ages that an hour consisted of exactly 22,560 indivisible "instants.") As we see here, the true nature of infinitesimals is quite different—they may be divided and divided again. Just as there is no smallest positive real, there is no smallest positive hyperreal.

There is no smallest among the small and no largest among the large; but always something still smaller and something still larger.
Anaxagoras (ca. 500B.C.–428B.C.)

must certainly be an infinitesimal. An obvious example is ◎/2. In general, we have the following fact:

THEOREM 4.1. If $◎_1$ and $◎_2$ are infinitesimals, and $r \neq 0$ is a real number, then

1. $◎_1 \cdot r$ is an infinitesimal;
2. $◎_1 \cdot ◎_2$ is an infinitesimal; and
3. if $◎_1 + ◎_2 \neq 0$, then it is an infinitesimal.

PROOF: We first see that $◎_1 \cdot r$ and $◎_1 \cdot ◎_2$ are both unequal to 0. This is because in the reals, if $a \neq 0$ and $b \neq 0$, then $a \cdot b \neq 0$. But we can write this as a sentence:

$$\forall x \, \forall y \, (x \neq 0 \land y \neq 0 \to xy \neq 0).$$

Since this sentence is true in the reals, it must also be true in HR. Thus, since r, $◎_1$, and $◎_2$ are all different from 0, $◎_1 \cdot r$ and $◎_1 \cdot ◎_2$ are different from 0.

Now all we need to show is that for any positive real s,

$$|◎_1 \cdot r| < s,$$
$$|◎_1 \cdot ◎_2| < s, \text{ and}$$
$$|◎_1 + ◎_2| < s.$$

We are again using the fact that **R** and HR satisfy the same sentences of L. For example,

But since $|◎_1| < s/|r|$, we get $|◎_1 \cdot r| < s$; and since $|◎_1| < s$ and $|◎_2| < 1$, we get $|◎_1 \cdot ◎_2| < s$; and finally, since $|◎_1| < s/2$ and $|◎_2| < s/2$, we get $|◎_1 + ◎_2| \leq |◎_1| + |◎_2| < s/2 + s/2 = s$. □

$$\forall x \forall y \forall z \left(|x| < \frac{y}{|z|} \rightarrow |x \cdot z| < y \right)$$

is a sentence true in **R** and hence in HR. This enables us to conclude that

$$|◎_1| < \frac{s}{|r|}$$

implies

$$|◎_1 \cdot r| < s.$$

Using the fact that the same sentences of L are true in both **R** and HR one can usually manipulate hyperreals exactly as one manipulates reals. If in doubt as to whether a manipulation, or property, of hyperreals is true, try to write out the appropriate sentence explicitly.

The symbol "□" is frequently placed at the end of proofs.

For any real number r, the set of hyperreals infinitely close to r was originally called the *monad* of r. This was in honor of Leibniz who had used the term "monad" in some of his (nonmathematical) philosophical works.

Hint: For problems 1 and 2, use theorem 4.1.

Hint: To prove that $|◎/r| < s$ for any two reals r and s, $s > 0$, show that $◎ < r \cdot s$.

Hint: Use theorem 4.1.

This theorem tells us that there are many infinitesimals, in fact, infinitely many. We could picture this as a cloud of infinitesimals surrounding the number 0.

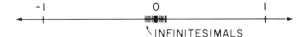

Are there other nonstandard numbers? For example, is there a nonstandard number between 4 and 5? One obvious candidate is $4 + ◎$. The general fact here is the following:

THEOREM 4.2. If r is real and h is nonstandard, then $r + h$ is nonstandard.

PROOF: Suppose not. Then for some real number, say s, we must have $r + h = s$. But this implies $h = s - r$. Since $s - r$ is real, h must now be a real, and this is a contradiction. □

This tells us that $4 + ◎$, $\pi - ◎$, $◎ + 2$, ... are all nonstandard. It also tells us that every real is surrounded by a cloud of nonstandard numbers "infinitely close" to it.

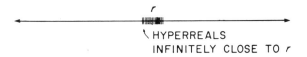

This is true because if $*$ is any infinitesimal around 0, then $r + *$ must be a nonstandard number "infinitely close" to r.

DEFINITION. Two hyperreals s and t are *infinitely close* if $s - t$ is an infinitesimal or zero. When s and t are infinitely close, we write

$$s \approx t.$$

EXERCISES
1. Prove that $◎ + ◎$ is nonstandard.
2. Prove that $◎^2$ is nonstandard.
3. Prove that if $r \neq 0$ is real, then $◎/r$ is an infinitesimal.
4. Prove that if $a < b < c$ and $a \approx c$, then $a \approx b$.
5. Prove that 4 and $4 + ◎$ are infinitely close.
6. Prove that if $a \approx b$ and $b \approx c$, then $a \approx c$.
7. Prove that if $a \approx b$ and $a \neq b$, then at least one of the two is nonstandard.

Hint: Let $\circledcirc_1 = a - b$, $\circledcirc_2 = c - d$; then fiddle around.

Hint: If a is not infinitesimal, $|a| > r_1 > 0$, and if b is not infinitesimal, $|b| > r_2 > 0$.

8. Prove that if $a \approx b$ and $c \approx d$, then $a + c \approx b + d$.
9. Prove that if $a \approx b$ and $c \approx d$, then $ac \approx bd$.
10. Prove that if ab is an infinitesimal, then either a or b is an infinitesimal.
11. Prove that $\sqrt{\circledcirc}$ is infinitesimal.

We know now that the hyperreal line is a lot denser than the real line, that is, more numbers are packed into the same space. Is the hyperreal line any *longer* than the real line? Is there a hyperreal number which is greater than all real numbers?

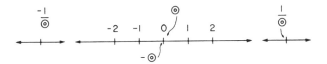

One obvious candidate is $1/\circledcirc$.

DEFINITION. A hyperreal is *infinite* if it is either greater than all real numbers or smaller than all real numbers. If a hyperreal is not infinite, then it is *finite*.

THEOREM 4.3. $1/\circledcirc$ is an infinite hyperreal.

PROOF: Since $\circledcirc > 0$, $1/\circledcirc > 0$, and so $1/\circledcirc$ is greater than all *negative* reals. If r is a positive real, then since $\circledcirc < 1/r$ it follows that $1/\circledcirc > r$; thus $1/\circledcirc$ is infinite. $\square$

It follows easily from this proof that if $*$ is any infinitesimal, then $1/*$ is infinite. Our picture of ℝ now looks like:

Once again our proof depends on our having a sentence,

$$\forall x_1 \forall x_2 \left(0 < x_1 < \frac{1}{x_2} \rightarrow \frac{1}{x_1} > x_2 \right),$$

that is true in **R**, hence in ℝ.

We admit, in geometry, not only infinite magnitudes, that is to say, magnitudes greater than any assignable magnitude, but infinite magnitudes infinitely greater, the one than other. This astonishes our dimension of brains, which is only about 6 inches long, 5 broad and 6 inches in depth in the largest heads.
Voltaire (1694–1778)

We now come to one of the great curiosities of the hyperreal numbers. You recall that one of our relations on the reals is $I(x)$, meaning "x is an integer," that is, $I(r)$ is true in the reals only if r is an integer. We already know that the hyperreals contain new *numbers*. We now ask the question: Do the hyperreals contain new *integers*? To be precise, is there a nonstandard number N such that $I(N)$ is true? The answer is yes.

THEOREM 4.4. ℝ contains infinite integers.

PROOF: We know that in the real numbers the following is true: Given any real number a, there is an integer n that is bigger.

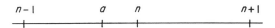

We can write this as a sentence in L:

$$\forall x_1 \, \exists x_2 (x_1 < x_2 \wedge I(x_2)).$$

Since this is true in the reals, it is true in the hyperreals. That means there must be a hyperreal number N bigger than $1/\odot$ such that $I(N)$ is true. N must be nonstandard since $1/\odot$ is bigger than all reals, hence N is an infinite integer. $\square$

In general when we wish to denote infinite integers, we will use the capital letters N, M, J, K, etc.

EXERCISES

1. Are there infinite irrational numbers?

2. Are there infinite prime numbers?

Hint: Let $Ir(x)$ mean "x is irrational," then proceed as in the proof of theorem 4.4.

3. Is there a smallest infinite integer?

4. If N_1, N_2 are infinite integers, is $N_1 + N_2$ an infinite integer?

5. If N_1, N_2 are infinite integers, is the greatest common divisor of N_1 and N_2 also an infinite integer?

Question: Are all *finite* integers real? The answer is yes, and you might try to prove it. If you have trouble, a proof is on the next page.

6. Are there infinite even integers? Odd integers?

7. Prove there is no nonstandard integer between 13 and 15.

DEFINITION. Let us say that if $0 < a < b$ and if b/a is infinite, then a is *infinitely smaller* than b, and b is *infinitely larger* than a.

When Jonathan Swift wrote the lines printed at the start of this chapter, he was thinking of calculus and, in particular, the idea of "infinitely smaller." The complete quotation is

8. Prove that $\odot$ is infinitely smaller than 1.

9. Prove that $\odot^2$ is infinitely smaller than $\odot$.

10. Prove that $\odot^3$ is infinitely smaller than $\odot^2$.

11. Prove that $1/\odot^2$ is infinitely larger than $1/\odot$.

So, Nat'ralists observe, a Flea Hath smaller Fleas that on him prey. And these have smaller Fleas to bite 'em. And so proceed ad infinitum.

12. Prove that $\sqrt{\odot}$ is infinitely larger than $\odot$ ($\odot > 0$).

13. Prove that if $a > 0$, then there is a number infinitely larger than a.

In the nineteenth century, the mathematician-logician Augustus De Morgan rewrote the lines of Swift and added another verse:

14. Prove that if $a > 0$, then there is a number infinitely smaller than a.

15. Is there a smallest positive infinite number?

16. Is there a largest infinitesimal?

Great Fleas have little fleas upon their backs to bite 'em
And the fleas have lesser fleas, and so ad infinitum.
And the great fleas themselves in turn have greater fleas to go on;
While these again have greater still, and so on.

17. Is there a smallest positive noninfinitesimal?

18. Is there a largest positive noninfinite number?

Answer to the question on page 37:
Suppose n is a finite integer. Let r be a real number such that $-r < n < r$. Let $n_1, n_2,..., n_k$ be the *real* integers between $-r$ and r. Then the following sentence is true in **R**: "For all numbers x_1, if x_1 is an integer between $-r$ and r, then x_1 is one of the numbers $n_1, n_2, ..., n_k$," that is,

$$\forall x_1[(I(x_1) \wedge -r < x_1 < r) \rightarrow (x_1 = n_1 \vee x_1 = n_2 \vee ... \vee x_1 = n_k)].$$

Since this is true in **R**, it is true in **HR** and so either $n = n_1$ or $n = n_2$ or ... or $n = n_k$. In any case, n is real.

The first conscientious use of decimal expansions was in 1585 by Simon Stevin.

Note: Whenever a number can be written with two decimal representations, for example

$$3 = 3.000000...$$
$$= 2.99999...,$$

we will use the representation without the 9s, that is, we will use

$$d(3, 1) = 0,$$
$$d(3, 2) = 0,$$
$$\vdots$$

rather than

$$d(3, 1) = 9,$$
$$d(3, 2) = 9,$$
$$\vdots$$

True or False? (answers on p. 40):
1. Between every two real numbers there is a nonstandard hyperreal number.
2. Between every two hyperreal numbers there is a real number.
3. The cube root of a nonstandard number is not necessarily nonstandard.

Often in dealing with real numbers, it is very useful to think of them in terms of decimal representations. Thus, it would be very nice if we could show that hyperreal numbers have decimal representations too. As it turns out, we can do this, and here is how.

Every decimal number has two parts, the integral part and the decimal part. For example,

$$s = 12{,}967. \underbrace{238165229...}$$
$$\underbrace{\phantom{12{,}967.}}_{\text{integral part}} \underbrace{}_{\text{decimal part}}$$

Let's define two functions to describe these:
1. $i(r) =$ the integral part of a given real r, in this case, $i(s) = 12{,}967$; other examples are $i(2\frac{1}{2}) = 2$, $i(\pi) = 3$.
2. $d(r, n) =$ the nth digit of the decimal part of a given real r, in this case,

$$d(s, 1) = 2$$
$$d(s, 2) = 3$$
$$d(s, 3) = 8$$
$$\vdots$$

Other examples are

$$d(2\tfrac{1}{3}, 1) = 3$$
$$d(2\tfrac{1}{3}, 2) = 3$$
$$d(2\tfrac{1}{3}, 3) = 3$$
$$\vdots$$

(since $2\frac{1}{3} = 2.333\bar{3} ...$);

$$d(\pi, 1) = 1$$
$$d(\pi, 2) = 4$$
$$d(\pi, 3) = 1$$
$$\vdots$$

(since $\pi = 3.14159 ...$).

Together, these two functions describe the desired representation of any real number r. In fact,

$$r = i(r).d(r, 1)d(r, 2)d(r, 3)d(r, 4) ....$$

What about hyperreal numbers? We still have the functions i and d defined, but here, since **HR** has more integers, $d(r, n)$ will be defined for n infinite as well as finite. Thus, the decimal representation of hyperreals is longer than that for reals. For example, if h is a

4. If $f(x)$ is real for all real numbers x, then $f(x)$ is real for all numbers x.

5. If $f(x) < 17$ for all real numbers x, then $f(x) < 17$ for all numbers x.

DEFINITION. We write $a \wr b$ if a and b are infinitely far apart, that is, $a - b$ is infinite.

6. $a \wr b$ implies $ac \wr bc$ for all c.

7. $a \wr b$ and $b \wr c$ imply $a \wr c$.

8. $a \wr b$ and $c \wr d$ imply $(a + c)$ $\wr (b + d)$.

9. $a \wr b$ and $a \neq 0$ and $b \neq 0$ imply $1/a \approx 1/b$.

Hint: The following sentence is true in the reals, hence also in the hyperreals:

$$\forall x_1((x_1 > 0 \wedge I(x_1)) \rightarrow d(\tfrac{1}{3}, x_1) = 3).$$

Hint: Let $E(x)$ mean "x is even," then consider the sentence:

$$\forall x_1((x_1 > 0 \wedge E(x_1)) \rightarrow d(\tfrac{3}{11}, x_1) = 7)$$

and so on.

Hint: First notice that if

$$0 < a < \frac{1}{10},$$

then $d(a, 1) = 0$. This can be written in a sentence of our language. Since it is true in **R**, it is true in HR, hence, since

$$0 < \text{\textcircled{\tiny ◎}} < \frac{1}{10},$$

$d(\text{\textcircled{\tiny ◎}}, 1) = 0$. Next, notice that if

$$0 < a < \frac{1}{100},$$

then $d(a, 2) = 0$, and so on.

given hyperreal, then

$i(h)$	exists,
$d(h, 1)$	exists,
$d(h, 2)$	exists,
$d(h, 3)$	exists, and so on,

but in addition, if N is any *infinite* integer, $N > 0$, then $d(h, N)$ exists too! So

$$h = i(h).d(h, 1)d(h, 2)d(h, 3)\ldots d(h, N)\ldots.$$

EXERCISES

1. We know that $d(\tfrac{1}{3}, 1) = 3$, $d(\tfrac{1}{3}, 2) = 3,\ldots$. If N is an infinite integer, what is $d(\tfrac{1}{3}, N)$?

2. We know that $d(3/11, 1) = 2$, $d(3/11, 2) = 7$, $d(3/11, 3) = 2$, $d(3/11, 4) = 7,\ldots$, that is, $3/11 = .272727\ldots$. If N is an infinite integer, what is $d(3/11, N)$?

3. What is $d(\text{\textcircled{\tiny ◎}}, 1)$, $d(\text{\textcircled{\tiny ◎}}, 2)$, $\ldots$?

There is one final question about the hyperreal line that we should answer: We know now that every real is infinitely close to many finite hyperreals, but is every finite hyperreal infinitely close to a real?

THEOREM 4.5. If h is any finite hyperreal, then there exists a standard real r infinitely close to h.

PROOF: Suppose we have some finite hyperreal h. Since h is finite, it lies somewhere between two *real* integers.

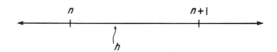

Then, just as for real numbers, $i(h)$ must be n. Let's look at h's decimal representation:

$$h = \underbrace{n.}_{\text{the real part}}\underbrace{d(h, 1)d(h, 2)d(h, 3) \ldots \ldots d(h, N)\ldots.}_{\substack{\text{decimal places} \\ N, \text{ where } N \text{ is} \\ \text{infinite}}}$$

The Hyperreal Line   39

The theorems we can prove about the hyperreal numbers are not restricted to ordinary functions, polynomials, etc. We can prove theorems about transcendental functions as well.

THEOREM. If ⊚ is an infinitesimal, then so is sin ⊚.

PROOF: For any x, if $0 < x < \pi/2$,

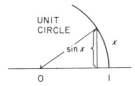

then $0 < \sin x < x$. Thus, if ⊚ > 0, $0 < \sin$ ⊚ $<$ ⊚. Similarly, if ⊚ < 0, ⊚ $< \sin$ ⊚ < 0. Hence, sin ⊚ is infinitesimal. □

EXERCISES
Assume ⊚ is an infinitesimal.
1. Prove $\cos$ ⊚ ≈ 1.
2. Prove $\sin(\pi/2 -$ ⊚$) \approx 1$.
3. Prove $\tan$ ⊚ ≈ 0.

If we retain only the real part

$$n.d(h, 1)d(h, 2)d(h, 3)...,$$

we have a real number r. It is a real number because it has an ordinary decimal representation. We claim that

$$r \approx h.$$

To see why this is true, we look at $h - r$. Since h and r have the same real part, that is

$$i(h) \quad = i(r)$$
$$d(h, 1) = d(r, 1)$$
$$d(h, 2) = d(r, 2)$$
$$\vdots$$

the decimal expansion of $h - r$ must start out all zeros,

$$i(h - r) \quad = 0$$
$$d(h - r, 1) = 0$$
$$d(h - r, 2) = 0$$
$$\vdots$$

This means that $h - r$ is infinitesimal or zero, since if s is any positive *real* number,

$$s > 0.0000 ...,$$

and so

$$s > h - r.$$

As $h - r$ is infinitesimal or zero, $h \approx r$. □

Notice that a hyperreal can be infinitely close to only *one* real. For, if $h \approx r_1$ and $h \approx r_2$, then $r_1 \approx r_2$ by exercise 6, p. 35. And so, by exercise 7, r_1 must equal r_2.

DEFINITION. If h is a finite hyperreal, let $\boxed{h}$ denote the unique real number infinitely close to h.

Now that we have our hyperreals we make one notational change. Instead of "⊚" standing for one particular infinitesimal, we will now use it as a variable standing for *any* infinitesimal.

EXERCISES
1. If h is finite and nonstandard, prove there is an infinitesimal ⊚ such that $\boxed{h} +$ ⊚ $= h$.
2. Prove that if r is real and ⊚ is an infinitesimal, then $\boxed{r + ⊚} = r$.
    Let h_1 and h_2 be finite hyperreals.
3. Prove $\boxed{h_1 + h_2} = \boxed{h_1} + \boxed{h_2}$.

As mentioned, the Axiom of Completeness (see p. 21) cannot be written as a sentence in the language L. Thus, even though it is true in **R**, it might not be true in HR. In fact, it isn't: Consider the set B of finite hyperreals. B is clearly nonempty and has an upper bound. Yet it has no least upper bound.

PROOF: Suppose q is a least upper bound. We will derive a contradiction. As q is an upper bound, q must be infinite (why?). But as q is infinite, so is $q - 1$, hence $q - 1$ is a *smaller* upper bound. Contradiction. $\square$

Here is one further computation we will find useful later:

$$\boxed{\frac{\sin \mathbb{O}}{\mathbb{O}}} = 1$$

for all infinitesimals $\mathbb{O}$.

PROOF: Consider the unit circle:

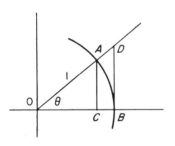

We have: The area of $\triangle OAC <$ the area of sector $OAB <$ the area of $\triangle ODB$, or

$$\frac{\cos \theta \sin \theta}{2} < \frac{\theta}{2} < \frac{\tan \theta}{2}.$$

Thus

$$\cos \theta < \frac{\theta}{\sin \theta} < \frac{1}{\cos \theta},$$

so for any infinitesimal $\mathbb{O} > 0$,

$$1 = \boxed{\cos \mathbb{O}} \leq \boxed{\frac{\mathbb{O}}{\sin \mathbb{O}}}$$

$$\leq \boxed{\frac{1}{\cos \mathbb{O}}} = 1.$$

Thus

$$\boxed{\frac{\mathbb{O}}{\sin \mathbb{O}}} = 1. \quad \square$$

4. Prove $\boxed{h_1 h_2} = \boxed{h_1} \cdot \boxed{h_2}$.
5. Prove $\boxed{h_1 - h_2} = \boxed{h_1} - \boxed{h_2}$.
6. Prove $\boxed{h_1 \div h_2} = \boxed{h_1 \div h_2}$ if $h_2 \neq 0$.
7. Prove that if $h_1 \leq h_2$, then $\boxed{h_1} \leq \boxed{h_2}$.
8. Prove that if $h \in [a, b]$, $a, b \in \mathbf{R}$ (a closed interval), then $\boxed{h} \in [a, b]$.

We now state and prove one final theorem. Despite its simplicity it is of fundamental importance and, beginning with the next chapter, is used again and again.

THEOREM 4.6. Let $a < b$ be real numbers. Then if p is any hyperreal in the interval $[a, b]$, so is $\boxed{p}$.

PROOF: Since p is in $[a, b]$, we must have $a \leq p \leq b$.

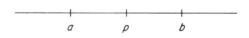

If $\boxed{p}$ were not in $[a, b]$, this would mean that either $\boxed{p} < a$ or $b < \boxed{p}$. Say $b < \boxed{p}$. Since $p \leq b < \boxed{p}$ and

$p \approx \boxed{p}$, we have $b \approx \boxed{p}$. But b and $\boxed{p}$ are distinct reals and hence cannot be infinitely close (or equal). This contradiction tells us $b < \boxed{p}$ is false. Similarly, so is $\boxed{p} < a$. Therefore $a \leq \boxed{p} \leq b$. $\square$

5

Continuous Functions

Any rigorous study of the calculus begins with one of continuous functions. They are the most fundamental objects of study for the analyst (as they are for most nonfinite mathematicians). It is they that turn approximation into precision and open the doors of calculus to reality.

Intuitively, a continuous function is one without gaps or holes. That is, not one like

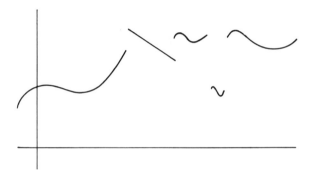

but rather one like

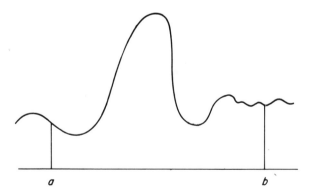

One obvious virtue of such an "unbroken" function is that it makes sense to ask the question: "What is the area bordered by the function and the lines $x = a$, $x = b$, and $y = 0$?"

If a function has gaps in it, the "area below" may be difficult or impossible to define.

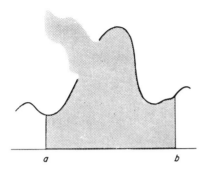

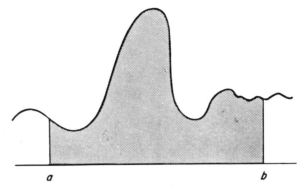

More will be said about this in the next chapter, but for now let's look at some theory concerning continuous functions themselves. The theorems we will prove are quite lovely and easily stand on their own, but don't let that fool you. They will be highly useful for our later work.

DEFINITION. A function f is said to be *continuous at* (a real number) r if the following condition is met: For every hyperreal p infinitely close to r, $f(p)$ is infinitely close to $f(r)$.

Intuitively, this definition should eliminate gaps, for if a function jumps

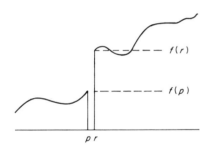

then we should be able to find $p \approx r$ such that $f(p) \not\approx f(r)$.

Symbolically,

f is continuous at r

iff

$p \approx r$ implies $f(p) \approx f(r)$.

DEFINITION. A function f is said to be *continuous* if it is continuous at *every* real number r.

EXERCISE
Why should p be of the form $1 + \bigcirc$?

An example: Is the function f given by $f(x) = x^2 + 2x - 1$ continuous at $x = 1$? Well, if p is infinitely close to 1, $p \neq 1$, then p is of the form $1 + \bigcirc$ where $\bigcirc$ is an infinitesimal. Therefore

$$f(p) = (1 + \bigcirc)^2 + 2(1 + \bigcirc) - 1 = 2 + \bigcirc^2 + 4\bigcirc.$$

But $f(1) = 2$, and so the difference between $f(1)$ and $f(p)$ is $\bigcirc^2 + 4\bigcirc$, something that is clearly infinitesimal (if you don't think it's so clear reread theorem 4.1).

Another example:

$f(x) = x - x^3$ at $x = -3$.

If p is infinitely close to -3, then $p = -3 + \bigcirc$ for some infinitesimal $\bigcirc$. Thus

EXERCISES
1. Prove $y = 4x^2 - 2x + 1$ continuous at $x = 3$.
2. Prove $y = 1/x$ continuous at $x = 1$.
3. Prove $y = \cos x$ continuous at any real point a.
4. Prove $y = x \sin x + 2$ continuous at $x = -1$.

Continuous Functions 43

$$f(p) = (-3 + \textcircled{\circ}) - (-3 + \textcircled{\circ})^3$$
$$= -3 + \textcircled{\circ} - (-27 + 9\textcircled{\circ}$$
$$- 3\textcircled{\circ}^2 + \textcircled{\circ}^3)$$
$$= 24 - 8\textcircled{\circ} + 3\textcircled{\circ}^2 - \textcircled{\circ}^3$$

and so $f(p) \approx 24 = f(-3)$.

Let us show that the function $\sin x$ is continuous at any real point a. If $p \approx a, p \neq a$, then $p = a + \textcircled{\circ}$ for some infinitesimal $\textcircled{\circ}$. Thus

$$\sin a - \sin p$$
$$= \sin a - \sin (a + \textcircled{\circ})$$
$$= \sin a - (\sin a \cos \textcircled{\circ}$$
$$+ \sin \textcircled{\circ} \cos a)$$
$$= \sin a (1 - \cos \textcircled{\circ})$$
$$+ \sin \textcircled{\circ} \cos a.$$

From the margin on page 40, $1 - \cos \textcircled{\circ}$ and $\sin \textcircled{\circ}$ are both infinitesimal, hence $\sin a - \sin p$ is infinitesimal or 0, so $\sin x$ is continuous at a.

Note: If $f(x)$ is *not* continuous at r, then we say f is *discontinuous* at r.

The following is an example of a *discontinuous* function:

$$f(x) = \begin{cases} 1 & \text{if } x \leq 0 \\ 2 & \text{if } x > 0. \end{cases}$$

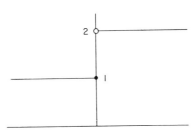

f is discontinuous at 0, since if $\textcircled{\circ}$ is a positive infinitesimal, $\textcircled{\circ} \approx 0$ but $f(\textcircled{\circ}) \not\approx f(0)$.

5. Prove $y = \cos (x^2 - x)$ continuous at $x = 3$.
6. Prove $y = \tan x$ continuous at $x = 2$.
7. Prove $y = \sqrt{x}$ continuous at any point $x > 0$.

As we mentioned earlier, our goal in the next chapter will be to compute the area below a positive, continuous function.

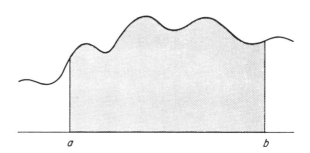

We have defined continuous such that this idea of "area" should make sense, but there are still some difficulties. One immediate problem is that we do not know that such an area is, in fact, finite. To be more specific, how can we be sure that a given continuous function does not grow so large between the lines $x = a$ and $x = b$ that it encompasses an infinite area?

THEOREM 5.1. Assume that f is a function continuous at every real number in the interval $[a, b]$. Then f is bounded on $[a, b]$, that is, there exists a real number h such that $-h < f(x) < h$ for every x in $[a, b]$.

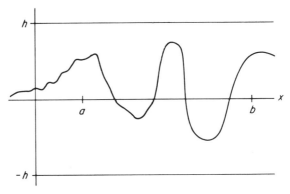

PROOF: We would like the sentence

$$(*) \quad \exists h \, \forall x (a \leq x \leq b \rightarrow -h < f(x) < h)$$

For this theorem we need two things: that f is continuous, and that our interval $[a, b]$ is closed and not open (a, b). For example, $f(x) = 1/x$ is continuous on $(0, 1)$, but not bounded. Prove this and find an example of a discontinuous function on $[1, 2]$ that is not bounded.

This theorem is sometimes called the "Extreme Value Theorem."

To the Greeks the ideal example of a continuous function was the path of a moving object, a ball, for example, or the point of a pen.

If we throw a ball up in the air and consider its path as a continuous function, this theorem states that there is a certain time when the ball is at its highest point.

Once again we need the closed interval and the continuous function even if we make the extra assumption that f is bounded. For example, $f(x) = x$ is bounded and continuous, but it does not assume a maximum on $(0, 1)$, since for any a in the interval, there is a b also in $(0, 1)$ such that $f(b) > f(a)$. Find a bounded, discontinuous function that assumes no maximum on $[1, 2]$.

to be true in the real numbers, and our method simply will be to show it true in the hyperreals. It is easy, however, to see that (∗) is true in HR for if h is any infinite (positive) hyperreal then

$$\forall x(a \le x \le b \to -h < f(x) < h).$$

This is because given *any* x in $[a, b]$, $\boxed{x}$ must be in $[a, b]$, and so the continuity of f tells us that $f(x) \approx f(\boxed{x})$. However, $f(\boxed{x})$ is a real number (hence finite) and so $f(x)$ is finite. As h is infinite, $-h < f(x) < h$. So $\exists h \, \forall x(a \le x \le b \to -h < f(x) < h)$ *is* true in the hyperreals (taking h to be any positive infinite)—it must thus also be true in the reals. □

There is a stronger result possible:

THEOREM 5.2. If f is a function continuous on the closed interval $[a, b]$, then not only is f bounded on $[a, b]$ but f actually assumes a maximum and a minimum on $[a, b]$, that is, there exist reals x_1 and x_2 in $[a, b]$ such that $f(x_1)$ is the maximum value f assumes on $[a, b]$ and $f(x_2)$ the minimum.

PROOF: We will show that f assumes a maximum—the proof for a minimum is similar.

Since f is bounded on $[a, b]$, let d be such that $-d < f(x) < d$ for every x in $[a, b]$.

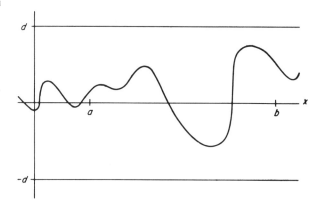

Now we know that if we divide the region between d and $-d$ into n equally spaced horizontal lines (for n any positive integer),

Note: The distance between lines is $2d/n$, so the mth line from the bottom is the line $y = -d + m(2d/n)$.

The reader may wish to compare what earlier mathematicians said about continuous functions and infinitesimals.

I do not consider mathematical quantities as consisting of the smallest possible parts, but as described by continuous motion.
Newton (1642–1727)

It is correct to consider quantities of which the difference is incomparably small to be equal.
Leibniz (1646–1716)

A quantity which is increased or decreased by an infinitesimally small quantity is neither increased nor decreased.
Johannes Bernoulli (1667–1748)

Let's consider another example of a discontinuous function:

$$f(x) = \begin{cases} \sin \dfrac{1}{x} & \text{if } x \neq 0 \\ 0 & \text{if } x = 0. \end{cases}$$

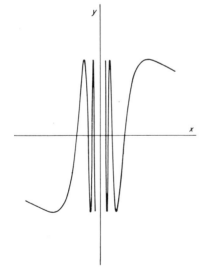

f is discontinuous at $x = 0$, for if ◎ $= 1/(2\pi N + \pi/2)$ where N is *infinite*, ◎ is certainly infinitesimal (hence infinitely close to 0), yet $f(◎)$ $= \sin (2\pi N + \pi/2) = 1$. Since 1 is not infinitely close to (or equal to)

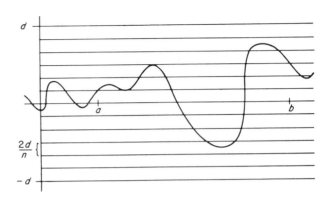

there must be a highest line above which f wanders. To say this another way: "For all n, if n is a positive integer, there is an m and a number $x \in [a, b]$ such that $f(x)$ is *above* the mth line (that is, $-d + m(2d/n) \leq f(x)$) and such that for all y, if $y \in [a, b]$, then $f(y)$ is *below* the $m + 1$st line (that is, $f(y) \leq -d + (m + 1) \cdot (2d/n)$)."

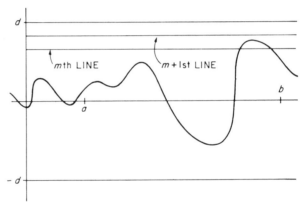

To say this yet another (and all important) way, the sentence

$$\forall n\Big((I(n) \wedge n > 0) \to \exists\, m\, \exists\, x\Big(I(m) \wedge 0 \leq m \leq n \,\wedge$$

$$a \leq x \leq b \wedge -d + m\Big(\frac{2d}{n}\Big) \leq f(x) \wedge \forall y\Big(a \leq y \leq b$$

$$\to f(y) \leq -d + (m + 1)\Big(\frac{2d}{n}\Big)\Big)\Big)\Big)$$

is true in the real numbers.

Now, as this sentence is true in the real numbers, and since it is written in our language L, it must also be true in HR, and so taking N to be an infinite positive hyperinteger, we see that there must be some hyperinteger M such that the Mth line is the highest above

0, we must conclude that $f(x)$ is not continuous at 0.

EXERCISES

Try to show that no matter how one redefines $f(x)$ at $x = 0$, the result is still discontinuous at 0.

Show that $f(x)$ *is* continuous at any $x \neq 0$.

How about the function

$$g(x) = \begin{cases} x \sin \dfrac{1}{x} & \text{if } x \neq 0 \\ 0 & \text{if } x = 0. \end{cases}$$

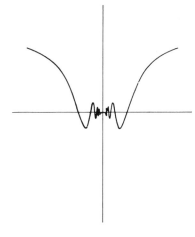

Is it continuous at 0? Yes. If ◎ is a given infinitesimal, we have to show ◎ sin 1/◎ infinitesimal or 0. Now, as ◎ is infinitesimal, ◎ sin 1/◎ would be infinitesimal or 0 if we could just show that sin 1/◎ is finite (even though 1/◎ is infinite). But notice that

$$\forall\, x(-1 \leq \sin x \leq 1)$$

is true in the reals, and hence it is true in the hyperreals. Thus

$$1 \leq \sin \frac{1}{◎} \leq 1.$$

This theorem is usually called the "Intermediate Value Theorem," and the property of assuming all values between $f(a)$ and $f(b)$ is called the "Darboux Property" of the function f.

which f wanders, that is, there is a hyperreal x such that

$$a \leq x \leq b \wedge -d + M\left(\frac{2d}{N}\right) \leq f(x) \wedge \forall y\Big(a \leq y \leq$$
$$b \to f(y) \leq -d + (M + 1)\left(\frac{2d}{N}\right)\Big).$$

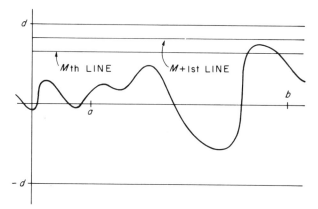

Note that

$$-d + M\left(\frac{2d}{N}\right) \leq f(x) \leq -d + (M + 1)\left(\frac{2d}{N}\right)$$

and thus

$$f(x) \approx -d + (M + 1)\left(\frac{2d}{N}\right)$$

since

$$-d + M\left(\frac{2d}{N}\right) \approx -d + (M + 1)\left(\frac{2d}{N}\right).$$

By our lemma $\boxed{x}$ is a real number in $[a, b]$, and we claim that $f(\boxed{x})$ is the maximum value f assumes on the reals in $[a, b]$. For as $f(\boxed{x})$ is infinitely close to $f(x)$, it is infinitely close to $-d + (M + 1)\,(2d/N)$. But any value of f on $[a, b]$ is at most $-d + (M + 1)\cdot(2d/N)$, and so no *real* value of f can exceed $f(\boxed{x})$. □

Our intuitive idea of a continuous function was a function with no "gaps" in it. It is now time that we vindicate our rigorous definition of continuous functions by showing that, indeed, they have no gaps.

THEOREM 5.3. Assume that f is a function continuous on $[a, b]$ and that r is any real number between $f(a)$ and $f(b)$. Then there is a real number x between a and b such that

$$f(x) = r.$$

Once again the continuity of f is very important. Find a discontinuous function f such that $f(1) < 0$, $f(2) > 0$, but f never equals 0 on $[1, 2]$.

If we consider the path of the point of a pencil to be a continuous function, then this theorem says that it is impossible to draw a line from the top of this page to the bottom without crossing this line:

PROOF: We will prove this for the case when $f(a) < r < f(b)$. The case when $f(a) > r > f(b)$ is nearly identical.

We will try to find our x by zeroing in on the place where f changes from being below r to above r.

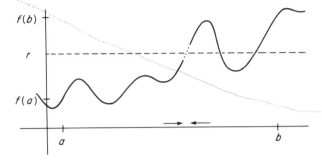

The following function is discontinuous at every real number:

$$f(x) = \begin{cases} 1 & \text{if } x \text{ is rational} \\ -1 & \text{if } x \text{ is irrational.} \end{cases}$$

Suppose r is any real number. We wish to prove that f is not continuous at r. Say, for example, r is rational. Let Rat (x) and Irr (x) be the relations "x is rational" and "x is irrational." Then we may write in L the sentence saying "between any two numbers lies an irrational" by

$\forall a \forall b (a < b \rightarrow \exists c (a < c < b \wedge \text{Irr} (c)))$.

This is true in $\mathbf{R}$, hence also in HR.

Thus, if $p \approx r$, we can find q between p and r such that q is irrational. Then we have $q \approx r$ but $-1 = f(q) \not\approx f(r) = 1$, so f is not continuous at r. Similarly, f is not continuous at any irrational point.

We know that if we divide $[a, b]$ into n pieces,

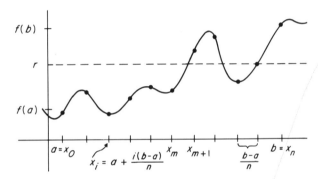

there must be two *adjacent* points x_m and x_{m+1} such that the function is below r at the first point and above it at the second. That is, "for any n, if n is a positive integer, there is an integer m between 0 and n such that $f(x_m) \leq r$ and $f(x_{m+1}) \geq r$."

To say this yet another way, the sentence

$$\forall \, n \Big((I(n) \wedge n > 0) \rightarrow \exists \, m \Big(I(m) \wedge 0 \leq m < n \wedge$$
$$f\Big(a + \frac{m(b-a)}{n}\Big) \leq r \wedge f\Big(a + \frac{(m+1)(b-a)}{n}\Big) \geq r \Big) \Big)$$

is true in $\mathbf{R}$. Since this is a sentence of L, its being true in $\mathbf{R}$ implies that it must also be true in HR, and so let us see what happens when our positive integer n is infinite. Suppose N is a positive infinite hyperinteger. Then our sentence tells us that there is a hyperinteger

EXERCISE
Prove theorem 5.2 by chopping $[a, b]$
up into many pieces as we did here.

(*Hint*: For some $M < N$,

$$f\left(a + \frac{M(b - a)}{N}\right)$$

is maximum.) (This is a hard
problem.)

Here is a neat proof that there
exists a root of the equation

$$x^5 - 3x^4 + 2x^3 - x^2 + x + 2 = 0.$$

The function

$$f(x) = x^5 - 3x^4 + 2x^2 - x^2 + x + 2$$

is continuous everywhere and hence,
in particular, on the interval $[-1, 1]$.
But $f(-1) = -6$ and $f(1) = 2$ and,
since 0 is between -6 and 2, there
must, by the Intermediate Value
Theorem, be some x_0 such that $f(x_0)$
$= 0$. x_0 is our desired root.

EXERCISES
1. Prove that every real number has a
cube root.
2. Prove all parabolas are continuous
(a parabola is a function of the form
$f(x) = ax^2 + bx + c$ for some real
numbers a, b, c).

Equivalently, we could say: Every
monotone function with the Darboux
Property is continuous.

Note: The assumption that f is monotone is necessary, for the function

$$f(x) = \begin{cases} \sin \dfrac{1}{x} & \text{if } x \neq 0 \\ 0 & \text{if } x = 0 \end{cases}$$

has the Darboux Property yet is
clearly not continuous. (See p. 46.)

M such that

$$f(x_M) = f\left(a + \frac{M(b - a)}{N}\right) \leq r \leq f\left(a + \frac{(M+1)(b-a)}{N}\right)$$
$$= f(x_{M+1}).$$

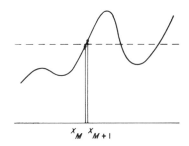

Let $x = \boxed{x_M}$, a real number. We claim that

$$f(x) = r.$$

For since

$$x_M \approx x_{M+1},$$

both $x \approx x_M$ and $x \approx x_{M+1}$ are true, hence

$$f(x) \approx f(x_M) \quad \text{and} \quad f(x) \approx f(x_{M+1}).$$

Since $f(x_M) \leq r \leq f(x_{M+1})$, $r \approx f(x)$, and since r and
$f(x)$ are both reals, $r = f(x)$. $\square$

There is a converse to this theorem. It isn't quite the
theorem that every function without gaps is continuous, but it's close. First, a definition:

DEFINITION. A function f is *monotone* on the interval
$[a, b]$ if either
1. for all $x, y \in [a, b]$, if $x < y$, then $f(x) \leq f(y)$; or
2. for all $x, y \in [a, b]$, if $x < y$, then $f(x) \geq f(y)$.

In case (1) we say f is *nondecreasing* on $[a, b]$, and in
case (2) we say f is *nonincreasing* on $[a, b]$.

THEOREM 5.4. Assume that f is monotone on $[a, b]$
and that for any r between $f(a)$ and $f(b)$ there is a
$c \in [a, b]$ such that $f(c) = r$. Then f is continuous on
$[a, b]$.

PROOF: Suppose f is *not* continuous on $[a, b]$ and that x
is a real and p a hyperreal such that

$$x \approx p$$

yet

The following is a particularly strange function:

$$f(x) = \begin{cases} 0 & \text{if } x \text{ is irrational} \\ \dfrac{1}{n} & \text{if } x \text{ is rational and } m/n \text{ represents } x \text{ in the simplest fractional form.} \end{cases}$$

First, f is discontinuous at all rational numbers.

PROOF: Suppose $x = m/n$ is a rational number, so that $f(x) = 1/n$. We know, as before, that between any two real numbers there is an irrational, so if we choose $p > m/n$, infinitely close to m/n, there is a q, $m/n < q < p$, that is irrational. We then have $m/n \approx q$, yet

$$f\left(\frac{m}{n}\right) = \frac{1}{n} \not\approx 0 = f(q).$$

Second, f is continuous at all irrational numbers!

PROOF: Suppose x is irrational so that $f(x) = 0$, and suppose $p \approx x$. Then either p is irrational or rational. If p is irrational, $f(p) = 0$ so $f(p) \approx f(x)$. Now, if p is rational, $p = M/N$ for some integers M and N, and $f(p) = 1/N$.

  Claim: N must be infinite, for if it is not, M must also be finite (since p is finite and $M = pN$) and so p is a real number.

  This is impossible because $p \neq x$, $p \approx x$.

  Finally, since N is infinite,

$$f(p) = \frac{1}{N} \approx 0 = f(x).$$

EXERCISES

1. Prove $f(x) = x$ is continuous at all points.
2. For any real r, prove $f(x) = r$ is continuous at all points.
3. Use exercises 1 and 2 and theorem 5.5 to prove that all polynomials are continuous. (*Hint*: Use induction.)

$f(x) \not\approx f(p).$

Since $f(x)$ and $f(p)$ are not infinitely close, there must be a real number r between them.

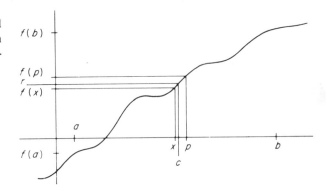

Since f is monotone, r is also between $f(a)$ and $f(b)$, hence there is a real number $c \in [a, b]$ such that $f(c) = r$. Again by monotonicity, c must lie between x and p. This, however, contradicts the fact that

$$x \approx p$$

(since x is real, there can be no reals between x and p). Thus, f is continuous. $\square$

We have already seen a number of examples of continuous functions. It should even be fairly easy for you to verify that every polynomial function is continuous. We now close this chapter with a general theorem about the existence of continuous functions. Specifically the theorem tells us how to get new continuous functions from old ones, and it is of great use.

THEOREM 5.5. Assume that f and g are functions continuous at r. Then the functions $f + g$, $f - g$, and fg are continuous at r. Furthermore, if $g(r) \neq 0$, then $1/g$ is continuous at r.

PROOF: Let p be any hyperreal infinitely close to r, and let $\bigcirc_1 = f(r) - f(p)$ and $\bigcirc_2 = g(r) - g(p)$. By continuity, $\bigcirc_1$ and $\bigcirc_2$ are either infinitesimal or 0. We must show that

1. $(f + g)(r) - (f + g)(p)$,
2. $(f - g)(r) - (f - g)(p)$,
3. $fg(r) - fg(p)$, and

THEOREM. If $y = f(x)$ and $z = g(y)$ are continuous functions, $z = g(f(x))$ is a continuous function.

PROOF: Exercise. (*Hint*: This is very simple.)

EXERCISE
Why is $\bigcirc_1 g(p)$ infinitesimal or 0?

EXERCISE
Why is

$$\frac{\bigcirc_2}{g(r)g(p)}$$

infinitesimal or 0?

EXERCISE
Let

$$f(p) = \begin{cases} x & \text{if } x \text{ is rational} \\ -x & \text{if } x \text{ is irrational.} \end{cases}$$

Prove that f is continuous only at $x = 0$. (*Hint*: See p. 48.)

Compare this with your ordinary idea of an interval.

Hint: Use theorem 5.3.

Hint: Use previous hint.

4. $1/g(r) - 1/g(p)$ (if $g(r) \neq 0$)
are all infinitesimal or 0. In (1) we have

$$f(r) + g(r) - f(p) - g(p) = \bigcirc_1 + \bigcirc_2$$

which is infinitesimal or 0. In (2) we have

$$f(r) - g(r) - f(p) + g(p) = \bigcirc_1 - \bigcirc_2$$

which is infinitesimal or 0. In (3) we have

$$\begin{aligned} f(r)g(r) &- f(p)g(p) \\ &= f(r)g(r) - f(r)g(p) + f(r)g(p) - f(p)g(p) \\ &= f(r)\bigcirc_2 + \bigcirc_1 g(p) \end{aligned}$$

which is infinitesimal or 0. Finally, in (4) we have

$$\frac{1}{g(r)} - \frac{1}{g(p)} = \frac{g(p) - g(r)}{g(r)g(p)} = \frac{\bigcirc_2}{g(r)g(p)}.$$

Assuming now that $g(r) \neq 0$, $g(p)$ must be infinitely close to $g(r)$. Thus $g(r)g(p)$ is a finite hyperreal, and hence the quotient

$$\frac{\bigcirc_2}{g(r)g(p)}$$

is infinitesimal or 0.   □

EXERCISES
1. Prove that $y = (x^2 - 1)/(x^2 + 1)$ is continuous at all points.
2. Prove that $y = |x|$ is continuous at all points.
3. Prove that if $f(x)$ and $g(x)$ are continuous at r and $g(r) \neq 0$, then $f(x)/g(x)$ is continuous at r.

DEFINITION. A set $X \subseteq \mathbf{R}$ is called an *interval* iff for all $a, b \in X$, if $a < c < b$, then $c \in X$.

4. Prove that if X is an interval and f is a continuous function, the range of f on X is an interval. (*Note*: The "range of f on X" is the set $\{ f(r) | r \in X \}$.)
5. Prove that if f is continuous on $[a, b]$ and $f(x)$ is real for all hyperreal x in $[a, b]$, then f is constant.

Continuous Functions

51

6

[Calculus:] The art of numbering and measuring a Thing whose existence cannot be conceived.
François Marie Arouet de Voltaire
(1694–1778)

Integral Calculus

The central problem of integral calculus is to determine the area of regions bounded by curves. It was assumed by the Greeks that every such region had an area, and great efforts were made to solve the problem for very specific areas. Within the last hundred years mathematicians have discovered that if the idea of region is generalized to mean any bounded subset of the plane, then it is not true that every region has an area. In this chapter we will consider one type of region, the area below a positive continuous function on a closed interval.

Any reasonable definition of "area" must satisfy certain obvious properties. For example, that the area of a rectangle is base times height or the area of congruent regions is the same. About one hundred years ago Van Vleck proved that there are bounded regions whose area cannot be defined so as still to satisfy our intuitive concepts.

The most dramatic example of this is the so-called Banach-Tarski paradox (Banach-analyst, Tarski-logician), which describes a method for chopping up a solid sphere into ten pieces and then reassembling them into two spheres the same size as the original sphere. Certainly these pieces cannot be considered to have a definite volume in any reasonable sense.

Measuring the size of bounded sets became an important problem in mathematics. Perhaps the most significant mathematical development of the early twentieth century was Lebesgue's theorem proving that a large number of sets do have a unique area.

Incidentally, it has also been shown that under some strange hypotheses, all bounded sets can be assigned a numerical "area." The universe under these hypotheses is a very odd and interesting one, but we are not interested in it here, for in that universe the hyperreal numbers do not exist!

As we mentioned before, some functions were integrated by the Greeks. In particular, the function $f(x) = x^2$ (see Chapter 1). This required finding, as Archimedes did,

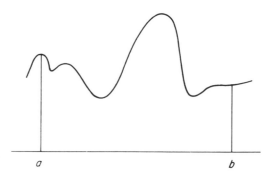

We shall prove that all such regions have an area.

Let's start with a simple problem: If $y = f(x)$ is a constant,

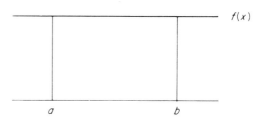

we know that the given region has an area, namely "base times height," that is, $(b - a) \cdot f(x)$. Of course, this is a trivial example but in a sense it is our basis for finding arbitrary areas. Indeed, if we think of an "arbitrary" region as consisting of rectangles standing next to one another,

the formula for $\sum_{i=1}^{k} i^2$, the sum of the first k squares. As we shall see, computing the integral of the function $f(x) = x^n$ by the methods of this chapter requires knowing the formula for $\sum_{i=1}^{k} i^n$. This proved something of a stumbling block for medieval and Renaissance mathematicians.

Another Greek, Nicomachus, found the formula for $\sum_{i=1}^{k} i^3$, and early Arabic mathematicians found $\sum_{i=1}^{k} i^4$. Later, the Arab Alhazon, ca. 965–1039 (also known as Ibn-al-Haitham), found a general method for finding the formula for $\sum_{i=1}^{k} i^n$. Using this, or something like it, Cavalieri computed

$$\int_0^1 x^n \, dx = \frac{1}{n+1}$$

for $n = 3, 4, \ldots, 9$.

Apparently by the time he got to $n = 10$, computational complexities overcame him. Not long after, the great mathematician Fermat computed the rest.

Considering their importance, let us examine the formulas for these sums.

The first, $\sum_{i=1}^{n} i = 1 + 2 + \cdots + n$, is particularly easy. We begin by representing the sum as a sum of boxes:

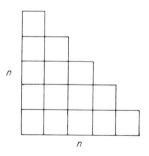

Putting two of these together,

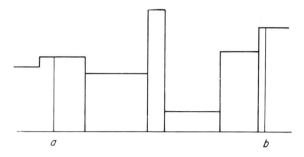

the area of any such region is simply the sum of the areas of the component rectangles. What about regions whose bounding function is very "curvy"?

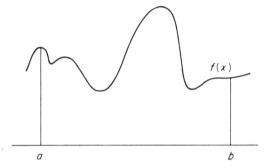

Simply think of it as composed of rectangles whose bases are infinitesimally small!

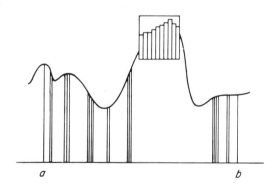

Now how can we make this formal? Let's look at the finite case first.

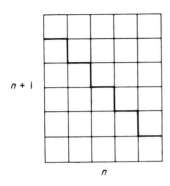

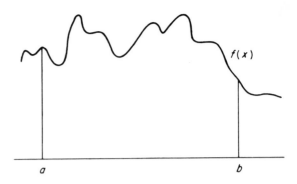

we get a rectangle with $n(n + 1)$ squares. Since this is exactly twice our desired sum, we have

$$\sum_{i=1}^{n} i = \frac{n(n + 1)}{2}.$$

To find the sum of the first n squares we use cubes. The sum $\sum_{i=1}^{n} i^2$ is represented by the volume of:

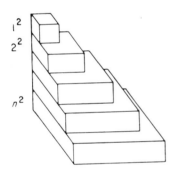

Adding another of these, we get:

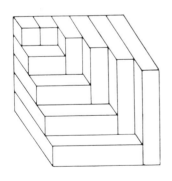

Given a function $f(x)$, an interval $[a, b]$, and a *real* distance $\Delta x > 0$, we can use Δx to "chop up" the interval $[a, b]$.

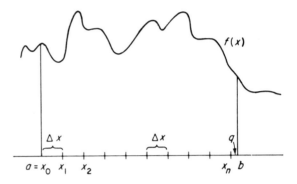

Since the length of each subdivision is Δx, we label the division points $x_0 (=a)$, x_1, x_2, x_3, ..., x_n where $x_i = a + i\Delta x$. If Δx divides $(b - a)$ evenly, then the last division point x_n will be b. If Δx does not divide $(b - a)$ evenly, there will be a small piece left over at the end. Let q be the remainder when $(b - a)$ is divided by Δx, so that

$$b - a = n\Delta x + q.$$

Adding a third, we get:

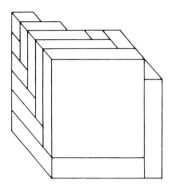

Finally, if we put another block of three, identical to this one, on top, we get:

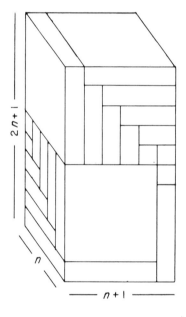

The volume of this is $n(n + 1) \cdot (2n + 1)$, and since this is six times the correct sum,

$$\sum_{i=1}^{n} i^2 = \frac{n(n + 1)(2n + 1)}{6}.$$

As we mentioned before, computing the integral

$$\int_0^1 x^n \, dx$$

quickly reduces to calculating the sum

$$\sum_{i=1}^{k} i^n.$$

At each little interval $[x_{i-1}, x_i]$ we draw a rectangle of height $f(x_i)$.

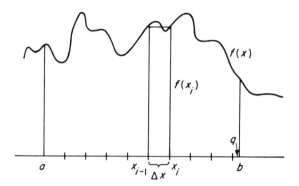

The area of the rectangle is $f(x_i) \, \Delta x$. We also draw a rectangle at the end to complete the picture.

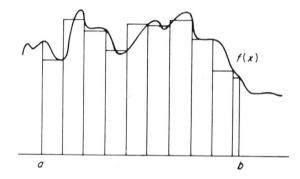

Notation: Let us use $\mathsf{S}_a^b f(x) \, \Delta x$ to denote the sum of the areas of the rectangles defined above. In other words,

$$\mathsf{S}_a^b f(x) \, \Delta x = \sum_{i=1}^{n} f(x_i) \, \Delta x + f(b) q.$$

This is an approximation of the area under $f(x)$ (and between a and b). To find an *infinitely close* approximation, we would like to find the sum $\mathsf{S}_a^b f(x) \, \Delta x$ when Δx, our rectangle base, is an infinitesimal. But why not just do it? Since $\mathsf{S}_a^b f(x) \, \Delta x$ can be viewed as a perfectly good function of the real variable Δx, a corresponding function symbol exists in our language L, *and so this function also becomes defined on the hyperreals*. Thus for any infinitesimal dx, the value of $\mathsf{S}_a^b f(x) (\, \cdot \,)$ at dx, $\mathsf{S}_a^b f(x) \, dx$, is a perfectly well-defined hyperreal.

Integral Calculus

This is really quite remarkable! $\int_0^1 x^n\,dx$ is a fundamentally infinite object whereas $\sum_i^k i^n$ is finite. Yet a thorough solution to the finite problem yields a solution to the related infinite one. This phenomenon appears throughout mathematics under the general heading *compactness*.

EXERCISES
Find the sum, in terms of n, of:
1. The first n even numbers, that is,

$$\sum_{i=1}^{n} 2i.$$

2. The first n odd numbers, that is,

$$\sum_{i=1}^{n} (2i - 1).$$

The term "integral" was first used by Jakob Bernoulli in 1690.
Notice that "x" in "$\int_a^b f(x)dx$" is a "dummy" variable. The value of the integral doesn't depend on any value of x. To put it another way,

$$\int_a^b f(q)dq = \int_a^b f(w)\,dw = \int_a^b f(\xi)\,d\xi.$$

Assuming that theorems 6.1 and 6.2 are true, let us attempt a few integrals:

Example 1. $f(x) = \frac{1}{2}x$:

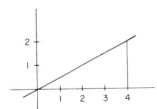

To test our method we will find the area of a triangle, that is, $\int_0^4 f(x)dx$. If our definition is any good at all, we should get $\frac{1}{2}4 \cdot 2 = 4$. Since $y = \frac{1}{2}x$ is continuous, theorem 6.2 says we may use any $dx > 0$. We will use $4/N$ for some infinite integer $N > 0$. *Note*: If $\varDelta x = 4/n$, $n > 0$, finite, then

$$\underset{0}{\overset{4}{S}} f(x)\varDelta x = \sum_{i=1}^{n} f(x_i)\frac{4}{n}$$

$$= \sum_{i=1}^{n} f\left(\frac{4i}{n}\right)\frac{4}{n}$$

DEFINITION. A function f defined on $[a, b]$ is said to be *integrable* if $\underset{a}{\overset{b}{S}} f(x)dx$ is finite and always has the same standard part for every positive infinitesimal dx.

DEFINITION. If $f(x)$ is integrable over $[a, b]$, we define the *integral of f over $[a, b]$* to be $\boxed{\underset{a}{\overset{b}{S}} f(x)dx}$ where dx is a positive infinitesimal. We denote the integral of f over $[a, b]$ by $\int_a^b f(x)dx$.

DEFINITION. If $f(x)$ is integrable and nonnegative on $[a, b]$, we define the *area under $f(x)$ between a and b* to be $\int_a^b f(x)dx$.

For these definitions to be useful, we would like many functions to be integrable. As it happens, every continuous function is integrable.

THEOREM 6.1. If f is continuous on $[a, b]$, then $\underset{a}{\overset{b}{S}} f(x)\,dx$ is finite for all infinitesimals $dx > 0$.

PROOF: By theorem 5.1, $f(x)$ is bounded, that is, there is a real $r > 0$ such that $-r < f(x) < r$ for all x in $[a, b]$. It is clear then, that for any real $\varDelta x$,

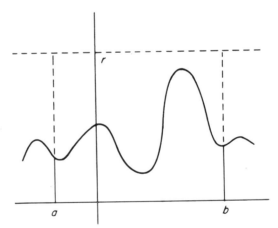

$$\underset{a}{\overset{b}{S}} f(x)\,\varDelta x \le r(b - a).$$

Similarly,

$$-r(b - a) \le \underset{a}{\overset{b}{S}} f(x)\,\varDelta x.$$

Since this is true for all real $\varDelta x$, it is true for all hyperreal dx:

$$-r(b - a) \le \underset{a}{\overset{b}{S}} f(x)\,dx \le r(b - a).$$

$$= \sum_{i=1}^{n} \frac{8i}{n^2}$$

$$= \frac{8}{n^2} \sum_{i=1}^{n} i.$$

From the previous discussion this equals

$$\frac{8}{n^2} \frac{n(n+1)}{2}.$$

Since this is true in **R** it is also true in **HR**, that is,

$$\overset{4}{\underset{0}{S}} f(x) \frac{4}{N} = \frac{8}{N^2} \frac{N(N+1)}{2}$$

$$= 4 + \frac{4}{N}.$$

Thus

$$\int_{0}^{4} f(x)\,dx = \boxed{4 + \frac{4}{N}} = 4.$$

Example 2. $f(x) = x^2 + x + 1$, from -1 to 2:

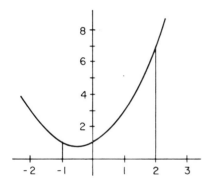

Again, letting $\varDelta x = 3/n$,

$$\overset{2}{\underset{-1}{S}} f(x)\,\varDelta x = \sum_{i=1}^{n} f(x_i) \frac{3}{n}$$

$$= \sum_{i=1}^{n} f\left(-1 + \frac{3i}{n}\right) \frac{3}{n}$$

$$= \sum_{i=1}^{n} \left(1 - \frac{6i}{n} + \frac{9i^2}{n^2} - 1 \right.$$

$$\left. + \frac{3i}{n} + 1\right) \frac{3}{n}$$

$$= \sum_{i=1}^{n} \left(\frac{27i^2}{n^3} - \frac{9i}{n^2} + \frac{3}{n}\right)$$

$$= \frac{27}{n^3} \sum_{i=1}^{n} i^2 - \frac{9}{n^2} \sum_{i=1}^{n} i$$

$$+ \frac{3}{n} \sum_{i=1}^{n} 1.$$

Thus, $\displaystyle \overset{b}{\underset{a}{S}} f(x)dx$ is finite for any infinitesimal $dx > 0$. $\square$

THEOREM 6.2. If f is continuous on $[a, b]$, then $\displaystyle \overset{b}{\underset{a}{S}} f(x)\,dx$ has the same standard part for all infinitesimals $dx > 0$.

PROOF: As always, we look at the finite case first. Given a function f,

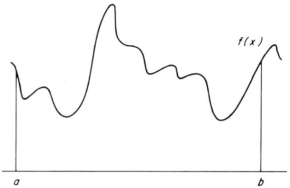

each $\varDelta x$ defines a new function.

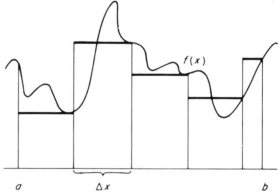

This new function is simply that which defines the tops of the rectangles of base $\varDelta x$ approximating the area under f between a and b.

Using formulas derived earlier, this equals

$$\frac{27}{n^3} \cdot \frac{n(n+1)(2n+1)}{6}$$

$$-\frac{9}{n^2} \cdot \frac{n(n+1)}{2} + \frac{3}{n} \cdot n$$

$$= 9 + \frac{27}{2n} + \frac{9}{2n^2} - \frac{9}{2} - \frac{9}{2n} + 3$$

$$= 7\tfrac{1}{2} + \frac{9}{n} + \frac{9}{2n^2}.$$

Thus

$$\underset{-1}{\overset{2}{S}} f(x)\frac{3}{N} = 7\tfrac{1}{2} + \frac{9}{N} + \frac{9}{2N^2}$$

and

$$\int_{-1}^{2} f(x)\,dx = \boxed{7\tfrac{1}{2} + \frac{9}{N} + \frac{9}{2N^2}}$$

$$= 7\tfrac{1}{2}.$$

EXERCISES

Compute the following integrals:

1. $\int_{0}^{2} 3x^2\, dx$.

2. $\int_{-2}^{-1} (x^2 - 4x + 2)\, dx$.

3. $\int_{0}^{4} \sqrt{x}\, dx$.

   Hint: Find the area of the shaded portion first:

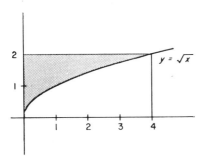

4. Prove that $\sum_{i=1}^{n} (h + 2ih^2 + i^2 h^3) = 7/3 + 3h/2 + h^2/6$ where $h = 1/n$ (see p. 11).

Here is an example of a function not integrable:

$$f(x) = \begin{cases} 1 & \text{if } x \text{ is rational} \\ 0 & \text{if } x \text{ is irrational.} \end{cases}$$

We will show that $\int_{0}^{1} f(x)\, dx$ doesn't exist. To see this, note that:

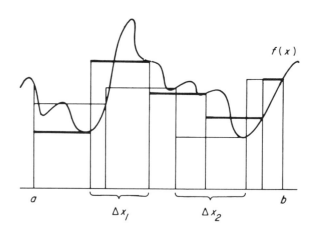

Let us call this new function $f_{\Delta x}$. It follows that

$$\underset{a}{\overset{b}{S}} f(x)\,\Delta x = \underset{a}{\overset{b}{S}} f_{\Delta x}(x)\,\Delta x.$$

Now, if we are given two *different* real numbers, Δx_1 and Δx_2, we will have two different rectangle approximations,

and hence we will have two different "step" functions $f_{\Delta x_1}$ and $f_{\Delta x_2}$.

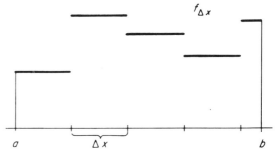

1. If Δx is rational, then each x_i is rational, so

$$\overset{1}{\underset{0}{S}} f(x) \, \Delta x = \sum_{i=1}^{n} f(x_i) \, \Delta x + f(1)q$$

$$= n \, \Delta x + q$$

$$= 1.$$

2. If Δx is irrational, then each x_i is irrational, so

$$\overset{1}{\underset{0}{S}} f(x) \, \Delta x = \sum_{i=2}^{n} f(x_i) \, \Delta x + f(1)q$$

$$= q.$$

Thus, if dx is a rational infinitesimal,

$$\left| \overset{1}{\underset{0}{S}} f(x) \, dx \right| = 1,$$

and if dx is an irrational infinitesimal,

$$\left| \overset{1}{\underset{0}{S}} f(x) \, dx \right| = 0,$$

since $0 < q < dx$. Since

$$\left| \overset{1}{\underset{0}{S}} f(x) \, dx \right|$$

is not independent of dx, $f(x)$ is not integrable over $[0, 1]$.

The key idea of our proof lies in the comparison of such step functions. Given Δx_1 and Δx_2, let diff(Δx_1, Δx_2) be the maximum difference between them,

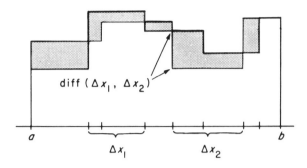

that is, let diff (Δx_1, Δx_2) equal the maximum value of $|f_{\Delta x_1}(x) - f_{\Delta x_2}(x)|$. Since

$$\left| \overset{b}{\underset{a}{S}} f_{\Delta x_1}(x) \, \Delta x_1 - \overset{b}{\underset{a}{S}} f_{\Delta x_2}(x) \, \Delta x_2 \right|$$

represents the shaded portion between the two step functions in the diagram above, it is easy to see that

$$\left| \overset{b}{\underset{a}{S}} f_{\Delta x_1}(x) \, \Delta x_1 - \overset{b}{\underset{a}{S}} f_{\Delta x_2}(x) \, \Delta x_2 \right|$$

$$\le (b - a) \operatorname{diff}(\Delta x_1, \Delta x_2),$$

so that

$$\left| \overset{b}{\underset{a}{S}} f(x) \, \Delta x_1 - \overset{b}{\underset{a}{S}} f(x) \, \Delta x_2 \right|$$

$$\le (b - a) \operatorname{diff}(\Delta x_1, \Delta x_2).$$

Since this is true for reals Δx_1 and Δx_2, it is also true for hyperreals dx_1 and dx_2:

$$\left| \overset{b}{\underset{a}{S}} f(x) \, dx_1 - \overset{b}{\underset{a}{S}} f(x) \, dx_2 \right| \le (b - a) \operatorname{diff}(dx_1, dx_2).$$

Let's now take a closer look at diff (dx_1, dx_2) and see if we can show that it's infinitesimal. First note that there are hyperreals c and d such that $c \approx d$,

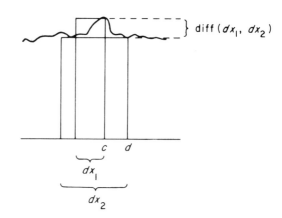

$\}$ diff (dx_1, dx_2)

c d

$\underbrace{\qquad}_{dx_1}$

$\underbrace{\qquad\qquad}_{dx_2}$

Why is $c \approx d$? Because the distance between c and d cannot be greater than $dx_1 + dx_2$.
Why is $e \in [a, b]$?

EXERCISE
Find the area bordered by the curve $f(x) = 9 + 6x - 3x^2$ and the x-axis. (*Hint*: Find where f crosses the x-axis.)

It is not true, however, that all integrable functions are continuous. To take a simple example, let

$$f(x) = \begin{cases} 1 & \text{if } x \neq 2 \\ 0 & \text{if } x = 2. \end{cases}$$

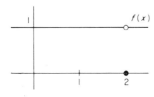

f is discontinuous at 2, since $f(2) = 0 \neq 1 = f(2 + \circledcirc)$ for any infinitesimal $\circledcirc$. On the other hand, for any Δx,

$$\overset{b}{\underset{a}{S}} f(x)\, \Delta x = \sum_{i=1}^{n} f(x_i)\, \Delta x + f(b)q.$$

If none of the x_i or b is the number 2, then

$$\overset{b}{\underset{a}{S}} f(x)\, \Delta x = \sum_{i=1}^{n} \Delta x + q$$

$$= n\, \Delta x + q$$

$$= b - a.$$

and max $(dx_1, dx_2) = f(c) - f(d)$. Let $e = \boxed{c}$. Then, as f is continuous on $[a, b]$, $f(e) \approx f(c)$ and $f(e) \approx f(d)$, and so $f(c) \approx f(d)$; max (dx_1, dx_2) is thus infinitesimal.

We are now virtually done, for

$$\left| \overset{b}{\underset{a}{S}} f(x)\, dx_1 - \overset{b}{\underset{a}{S}} f(x)\, dx_2 \right| \leq (b - a) \max (dx_1, dx_2)$$

is an infinitesimal, so

$$\boxed{\overset{b}{\underset{a}{S}} f(x)\, dx_1} = \boxed{\overset{b}{\underset{a}{S}} f(x)\, dx_2}.$$

Since dx_1 and dx_2 were arbitrary positive infinitesimals, our proof is complete. $\square$

By this theorem, our choice of dx is irrelevant. We may choose any dx we like for computing $\int_a^b f(x)dx$. In general, we will choose dx equal to $(b - a)/N$ (where N is a positive infinite integer), so that dx will divide $b - a$ evenly and there will be no remainder.

Note that the integral is defined for all continuous functions, not just positive ones. Although this has no direct application to areas, there will be uses for it later.

We can also define

$$\int_a^b f(x)\, dx \qquad \text{for } a > b.$$

DEFINITION. If f is integrable on $[b, a]$, where $b < a$, we define $\int_a^b f(x)\, dx$ to be $-\int_b^a f(x)\, dx$.

THEOREM 6.3. If f is continuous on $[d, e]$ and a, b, c are in $[d, e]$, then $\int_a^b f(x)\, dx = \int_a^c f(x)\, dx + \int_c^b f(x)\, dx$.

Otherwise, if one of the x_i equals 2, then

$$\overset{b}{\underset{a}{S}} f(x) \, \varDelta x = b - a - \varDelta x,$$

or if $b = 2$,

$$\overset{b}{\underset{a}{S}} f(x) \, \varDelta x = b - a - q.$$

In either case,

$$\left| \overset{b}{\underset{a}{S}} f(x) \, \varDelta x - (b - a) \right| \leq \varDelta x.$$

Thus, if dx is infinitesimal,

$$\left| \overset{b}{\underset{a}{S}} f(x) \, dx - (b - a) \right| \leq dx,$$

so

$$\boxed{\overset{b}{\underset{a}{S}} f(x) \, dx} = (b - a).$$

Thus, since $\left| \overset{b}{\underset{a}{S}} f(x) \, dx \right|$ is

always constant for different dx, f is integrable.

EXERCISES

1. Prove for any integrable functions f and g, and reals a, b, that

$$\int_a^b f(x) \, dx + \int_a^b g(x) \, dx$$

$$= \int_a^b [f(x) + g(x)] \, dx.$$

Hint: Show that for all $\varDelta x$,

$$\overset{b}{\underset{a}{S}} f(x) \, \varDelta x + \overset{b}{\underset{a}{S}} g(x) \, \varDelta x$$

$$= \overset{b}{\underset{a}{S}} [f(x) + g(x)] \, \varDelta x.$$

2. Prove for any integrable function $f(x)$, and any real numbers k, a, and b, that

$$k \int_a^b f(x) \, dx = \int_a^b k f(x) \, dx.$$

These two theorems were first proved by Cavalieri in the seventeenth century, although his proof of the second is considered shaky, even by seventeenth-century standards.

PROOF: There are six cases to consider:

(1) $a \leq c \leq b$
(2) $a \leq b \leq c$
(3) $b \leq a \leq c$
(4) $b \leq c \leq a$
(5) $c \leq b \leq a$
(6) $c \leq a \leq b$

We will do case (1) and leave the others as exercises. In terms of areas, case (1) states a familiar fact, that

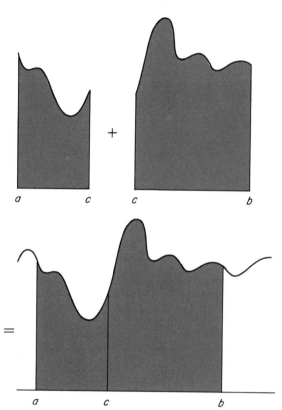

In Chapter 8 we will find still simpler proofs of the theorems.

Cavalieri's method used what he termed "indivisibles," quantities of not merely infinitesimal size, but actually of size 0. It was one of many fruitless attempts to justify infinitesimals. (See p. 34.)

(See p. 34.)

Let's proceed formally. Since f is continuous, we may choose any dx we like. We choose $dx = (c - a)/N$ for some infinite integer N. This is useful to do, for if n is finite and if $\Delta x = (c - a)/n$, then the diagram

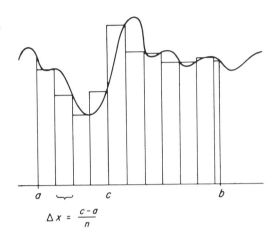

$$\Delta x = \frac{c-a}{n}$$

immediately shows that

$$\overset{b}{\underset{a}{S}} f(x) \, \Delta x = \overset{c}{\underset{a}{S}} f(x) \, \Delta x + \overset{b}{\underset{c}{S}} f(x) \, \Delta x.$$

Thus

$$\overset{b}{\underset{a}{S}} f(x) \, dx = \overset{c}{\underset{a}{S}} f(x) \, dx + \overset{b}{\underset{c}{S}} f(x) \, dx,$$

and so

$$\int_a^b f(x) \, dx = \int_a^c f(x) \, dx + \int_c^b f(x) \, dx. \quad \square$$

EXERCISE
Complete the proof of theorem 6.3. *Hint:* For case 2,

$$\int_a^c f(x) \, dx = \int_a^b f(x) \, dx + \int_b^c f(x) \, dx.$$

Thus

$$\int_a^c f(x) \, dx = \int_a^b f(x) \, dx - \int_c^b f(x) \, dx$$

or

$$\int_a^c f(x) \, dx + \int_c^b f(x) \, dx = \int_a^b f(x) \, dx.$$

We have been dealing throughout this chapter with the area under a curve. Let us now define the area function itself. Suppose f is continuous and suppose we define a function $F(x)$ by

$$F(x) = \int_c^x f(t) \, dt \qquad (c \text{ a constant}).$$

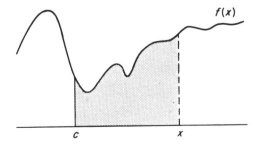

Then $F(x)$ has the property that for any a and b,

$$\int_a^b f(x)\,dx = F(b) - F(a). \tag{1}$$

This is true from theorem 6.3:

$$\begin{aligned}
F(b) - F(a) &= \int_c^b f(x)\,dx - \int_c^a f(x)\,dx \\
&= \int_c^b f(x)\,dx + \int_a^c f(x)\,dx \\
&= \int_a^b f(x)\,dx.
\end{aligned}$$

This is really quite remarkable. On one hand the area under the function f between a and b would certainly appear to intimately depend on the entire region between a and b. Yet on the other hand our formula (1)

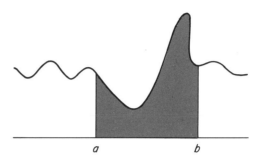

tells us that we can find this area by simply examining the behavior of some other function *at the endpoints* of the region, paying absolutely no regard to internal points.

DEFINITION. Given any continuous function g, a function H with the property that for all a and b,

$$\int_a^b g(t)\,dt = H(b) - H(a),$$

is called a *primitive* of g.

The following are some examples of primitives.

1. Consider $f(x) = k$, a constant:

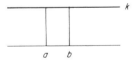

Since $\int_a^b k\,dx = k(b - a) = kb - ka$, $g(x) = kx$ is a primitive of f.

2. $f(x) = x^2$. Let's compute $\int_0^x f(t)\,dt$. Choosing $\varDelta t = x/n$,

$$\begin{aligned}
\mathop{S}_0^x f(t)\,\varDelta t &= \sum_{i=1}^n f\left(\frac{ix}{n}\right)\frac{x}{n} \\
&= \sum_{i=1}^n \frac{i^2 x^2}{n^2}\frac{x}{n} \\
&= \frac{x^3}{n^3}\sum_{i=1}^n i^2 \\
&= \frac{x^3}{n^3}\frac{n(n+1)(2n+1)}{6}.
\end{aligned}$$

As we saw earlier, given any continuous g and constant c, $G(x) = \int_c^x g(t)dt$ is a primitive of g, and although not every primitive of g is of the form $\int_c^x g(t)dt$, it almost is.

THEOREM 6.4. Let c be any real number and g any continuous function. Then H is a primitive of g iff for some constant k,

$$H(x) = \int_c^x g(t)\,dt + k.$$

PROOF: Let $G(x)$ be defined by

$$G(x) = \int_c^x g(t)\,dt.$$

So for $dx = x/N$,

$$\overset{x}{\underset{0}{S}} f(t)\,dt = x^3 \cdot \boxed{\dfrac{2N^3 + 3N^2 + N}{6N^3}}$$

$$= x^3 \cdot \boxed{\dfrac{1}{3} + \dfrac{1}{2N} + \dfrac{1}{6N^2}}$$

$$= \tfrac{1}{3}x^3.$$

Thus $g(x) = \tfrac{1}{3}x^3$ is a primitive of f.

EXERCISES

1. Find a primitive of $f(x) = x$.
2. Find a primitive of $f(x) = x^3$.
3. Find a primitive of $f(x) = \sqrt{x}$
(see p. 58).

As discussed earlier, $G(x)$ is a primitive of g. If, now, $H(x)$ is any function such that for some k

$$H(x) = G(x) + k,$$

then for any a and b,

$$H(b) - H(a) = G(b) + k - G(a) - k$$

$$= G(b) - G(a) = \int_a^b g(t)\,dt.$$

Thus H must be a primitive. Conversely, if $H(x)$ is any primitive of g,

$$H(x) - H(c) = \int_c^x g(t)\,dt = G(x).$$

Therefore $H(x) = G(x) + H(c)$. So if we let $k = H(c)$,

$$H(x) = G(x) + k. \quad \square$$

7

The theory of limits is the true metaphysics of the differential calculus.
Jean Le Rond d'Alembert
(1717–1783)

Differential Calculus

Jean d'Alembert was a French philosopher and mathematician. At the time that he made the statement above, calculus was only beginning to be understood. Mathematicians, unable at the time to put the infinitesimals of Leibniz on a firm foundation, began to use limits to rigorously base Newton's "ultimate ratios." Today, the problem of infinitesimals has been solved, and the modern nonstandard analyst would reply to d'Alembert: "*Mathematical logic* is the true metaphysics of the differential calculus."

Interestingly, it is possible to find the tangents to some curves without calculus. An example from Chapter 1: We can find the tangent to $y = x^2$ at $x = 1$.

The central problem of differential calculus is to determine tangents to curves. The Greeks, while successful for a number of specific curves, had no means of approaching the problem in general. In our development we will place the question in its clearest setting, the analytic geometry of Descartes, and attack it with the hyperreal numbers. Using techniques we have already established, the problem is not difficult at all.

The slope of a straight line is usually defined as $\Delta y/\Delta x$, and in high school algebra it is shown that this ratio is constant, regardless of the size of Δx.

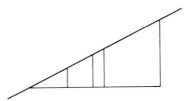

With curved lines this is not true.

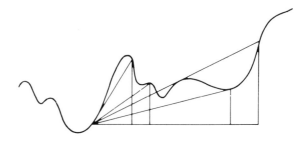

Intuitively, if we take Δx very small, then $\Delta y/\Delta x$ should approximate the true slope fairly well. If we take Δx to be infinitesimal, it should approximate the true slope infinitely well.

We are looking for a line $y = mx + a$ that intersects $y = x^2$ at exactly one point: $(1, 1)$. Since this point is on the line, we have

$$1 = m \cdot 1 + a,$$

so that

$a = 1 - m$.

At the intersection of the curve and the line, we have

$x^2 = mx + a = mx + 1 - m$

or

$x^2 - mx + m - 1 = 0$.

Using the quadratic formula:

$$x = \frac{m \pm \sqrt{m^2 - 4m + 4}}{2}$$

$$= \frac{m \pm (m - 2)}{2}$$

$$= m - 1 \text{ or } 1.$$

Since there is only one point of intersection, $m - 1 = 1$, therefore $m = 2$ and the tangent is $y = 2x - 1$.

Note: It is immediate from this definition that the derivative of a line $f(x) = mx + a$ at any point is its slope m, for

$$\left| \frac{f(b + \Delta x) - f(b)}{\Delta x} \right|$$

$$= \left| \frac{mb + m \Delta x + a - mb - a}{\Delta x} \right|$$

$$= m \qquad \text{for all } \Delta x.$$

Since $\Delta f / \Delta x \approx f'(b)$, $\Delta f \approx f'(b) \, \Delta x$.

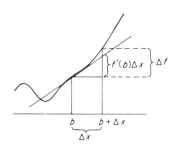

A few quick examples of derivatives:

1. $f(x) = x^2$ at $x = 1$:

$$\frac{\Delta f}{\Delta x} = \frac{(1 + \Delta x)^2 - 1^2}{\Delta x}$$

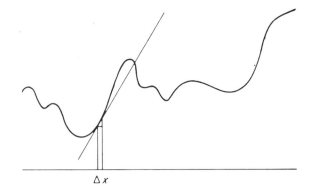

DEFINITION. If $y = f(x)$ is a function defined at $x = b$ and if

$$\left(\frac{\Delta f}{\Delta x} \right)_b = \frac{f(b + \Delta x) - f(b)}{\Delta x}$$

is finite and has the same standard part no matter which infinitesimal Δx we choose, then f is said to be *differentiable at b*, and the *derivative of f at b*, written $f'(b)$, is

$$f'(b) = \left| \frac{f(b + \Delta x) - f(b)}{\Delta x} \right| = \left| \frac{\Delta f}{\Delta x} \right|.$$

In earlier chapters we have dealt exclusively with continuous functions. It turns out that differentiable functions are continuous.

THEOREM 7.1. If $f(x)$ is differentiable at b, then it is continuous at b.

PROOF: For any infinitesimal Δx, $(f(b + \Delta x) - f(b))/\Delta x$ is finite, and hence $f(b + \Delta x) - f(b)$ must be infinitesimal or zero. $\square$

The converse to this theorem is not true. The function $f(x) = |x|$ is clearly continuous at 0 yet is not differentiable there (see exercises 5, 6, p. 70). The situation then, with our properties differentiable, continuous, and integrable, is this: The first implies the second which implies the third, yet the third does not imply the second which does not imply the first.

DEFINITION. If f is differentiable at every point in an interval $[a, b]$, then f is *differentiable on $[a, b]$*. Note that if f is differentiable on $[a, b]$, then we have defined a new function on $[a, b]$, f'.

As the examples on the left demonstrate, it is certainly easier to differentiate functions than to integrate

$$= \frac{1 + 2\Delta x + \Delta x^2 - 1}{\Delta x}$$

$$= 2 + \Delta x,$$

so $\boxed{\Delta f / \Delta x} = 2$ and $f'(1) = 2$, the slope of the tangent as found on p. 66.

2. $f(x) = 2x^3 - 3x^2 + 6x - 1$ at $x = 4$:

$$\frac{\Delta f}{\Delta x} = \frac{\begin{aligned}&[2(4 + \Delta x)^3 - 3(4 + \Delta x)^2 \\ &+ 6(4 + \Delta x) - 1 - 2(4)^3 \\ &+ 3(4)^2 - 6(4) + 1]\end{aligned}}{\Delta x}$$

$$= \frac{\begin{aligned}&[128 + 96\Delta x + 24(\Delta x)^2 \\ &+ 2(\Delta x)^3 - 48 - 24\Delta x \\ &- 3(\Delta x)^2 + 24 + 6\Delta x - 1 \\ &- 128 + 48 - 24 + 1]\end{aligned}}{\Delta x}$$

$$= 78 + 21\Delta x + 2(\Delta x)^2.$$

Thus for any infinitesimal Δx,

$$\frac{\Delta f}{\Delta x} = \boxed{78 + 21\Delta x + 2(\Delta x)^2} = 78$$

and $f'(4) = 78$.

For some time Leibniz believed that

$$h'(b) = f'(b)g'(b),$$

which seemed more natural. This illustrates a point often forgotten: The easy theorems of today were the hard theorems of yesterday.

EXERCISE
Find an example of functions f and g such that if $h = fg$, then there is a b where

$$h'(b) \neq f'(b)g'(b).$$

One special case of theorem 7.3: If $f(x) = k$, a constant, then $f'(x) = 0$ (again, since the slope of the line $y = k$ is 0), so the derivative of $kg(x)$ is $kg'(x)$ by theorem 7.3.

them, although differentiating more complex functions can be fairly involved. The following four theorems are of enormous help:

THEOREM 7.2. If f and g are differentiable at b and $h(x) = f(x) + g(x)$, then h is differentiable at b and $h'(b) = f'(b) + g'(b)$.

PROOF: Given an infinitesimal Δx, let

$$\Delta h = h(b + \Delta x) - h(b),$$

$$\Delta f = f(b + \Delta x) - f(b),$$

and

$$\Delta g = g(b + \Delta x) - g(b).$$

Then

$$\boxed{\frac{\Delta h}{\Delta x}} = \boxed{\frac{\Delta f + \Delta g}{\Delta x}} = \boxed{\frac{\Delta f}{\Delta x}} + \boxed{\frac{\Delta g}{\Delta x}} = f'(b) + g'(b).$$

Thus $h'(b) = f'(b) + g'(b)$. $\square$

EXERCISES
Find $f'(a)$ where:
1. $f(x) = x^2$, $a = 3$.
2. $f(x) = 5 - x^2$, $a = -1$.
3. $f(x) = x^2$, $a = 2$.
4. $f(x) = 2x^4 - 3x^2 + 1$, $a = 0$.
5. Prove that if f and g are differentiable at b and

$$h(x) = f(x) - g(x),$$

then h is differentiable at b and $h'(b) = f'(b) - g'(b)$.

This next theorem is sort of a surprise. Since it is so unexpected, it took Leibniz several years to discover it.

THEOREM 7.3. If f and g are differentiable at b, and $h(x) = f(x)g(x)$, then h is differentiable at b and $h'(b) = f'(b)g(b) + f(b)g'(b)$.

PROOF: Given an infinitesmial Δx, let

$$\Delta f = f(b + \Delta x) - f(b),$$

$$\Delta g = g(b + \Delta x) - g(b),$$

and

$$\Delta h = h(b + \Delta x) - h(b).$$

Then

$$\boxed{\frac{\Delta h}{\Delta x}} = \boxed{\frac{(f(b) + \Delta f)(g(b) + \Delta g) - f(b)g(b)}{\Delta x}}$$

Another function differentiated: find the derivative of $f(x) = \sqrt{x}$ at any point $x > 0$.

$$\frac{\Delta f}{\Delta x} = \frac{\sqrt{x + \Delta x} - \sqrt{x}}{\Delta x}$$

$$= \frac{\sqrt{x + \Delta x} - \sqrt{x}}{\Delta x} \cdot$$

$$\frac{\sqrt{x + \Delta x} + \sqrt{x}}{\sqrt{x + \Delta x} + \sqrt{x}}$$

$$= \frac{x + \Delta x - x}{\Delta x(\sqrt{x + \Delta x} + \sqrt{x})}$$

$$= \frac{1}{\sqrt{x + \Delta x} + \sqrt{x}};$$

thus for any infinitesimal Δx,

$$\left| \frac{\Delta f}{\Delta x} \right| = \frac{1}{2\sqrt{x}}$$

since f is continuous (p. 45). Hence $f'(x) = 1/(2\sqrt{x})$.

If we had assumed f was differentiable, we could have computed $f'(x)$ using theorem 7.3, for if $h(x) = f(x) \cdot f(x) = x$, then $h'(x) = 1$ from page 66, and by theorem 7.3,

$$h'(x) = f(x)f'(x) + f'(x)f(x),$$

or

$$1 = 2\sqrt{x}f'(x).$$

Thus

$$f'(x) = 1/(2\sqrt{x}).$$

Why is $\overline{|f(b + \Delta x)|} = f(b)$? (*Hint:* Use theorem 7.1.)

EXERCISE
Find $f'(x)$ for all points x. (If necessary, use the theorems.)
1. $f(x) = x$.
2. $f(x) = x^2$.
3. $f(x) = 7x^2 - 2x + 3$.
4. $f(x) = x^2\sqrt{x}$.
5. $f(x) = 1/x$.
6. $f(x) = x^2/(3\sqrt{x})$.

$$= \overline{\frac{f(b)\Delta g + \Delta f g(b) + \Delta f \Delta g}{\Delta x}}$$

$$= \overline{|f(b)|} \, \overline{\left|\frac{\Delta g}{\Delta x}\right|} + \overline{\left|\frac{\Delta f}{\Delta x}\right|} \, \overline{|g(b)|} + \overline{|\Delta f|} \, \overline{\left|\frac{\Delta g}{\Delta x}\right|}$$

$$= f(b)g'(b) + f'(b)g(b) + 0,$$

so

$$h'(b) = f(b)g'(b) + f'(b)g(b). \quad \square$$

THEOREM 7.4. If f is differentiable at b and $f(b) \neq 0$, then the function $h(x) = 1/f(x)$ is also differentiable at b and

$$h'(b) = -\frac{f'(b)}{[f(b)]^2}.$$

PROOF: Given an infinitesimal Δx, let

$$\Delta f = f(b + \Delta x) - f(b)$$

and

$$\Delta h = h(b + \Delta x) - h(b).$$

Then

$$\overline{\left|\frac{\Delta h}{\Delta x}\right|} = \overline{\left|\frac{1}{\Delta x}\left(\frac{1}{f(b + \Delta x)} - \frac{1}{f(b)}\right)\right|}$$

$$= \overline{\left|\frac{1}{\Delta x}\left(-\frac{f(b + \Delta x) - f(b)}{f(b + \Delta x)f(b)}\right)\right|}$$

$$= \overline{\left|-\frac{\Delta f}{\Delta x}\frac{1}{f(b + \Delta x)f(b)}\right|}$$

$$= -\overline{\left|\frac{\Delta f}{\Delta x}\right|}\frac{1}{\overline{|f(b + \Delta x)|}\,\overline{|f(b)|}}$$

$$= -\frac{f'(b)}{[f(b)]^2}.$$

Thus

$$h'(b) = \frac{-f'(b)}{[f(b)]^2}. \quad \square$$

Theorems 7.3 and 7.4 together yield:

COROLLARY. If f and g are differentiable at b and $g(b) \neq 0$, then the function $h(x) = f(x)/g(x)$ is differentiable at b and

$$h'(b) = \frac{f'(b)g(b) - g'(b)f(b)}{[g(b)]^2}.$$

Another example: sin x. To find the derivative of sin x at $x = a$, let Δx be any infinitesimal. Then

$$\boxed{\frac{\sin(a + \Delta x) - \sin a}{\Delta x}}$$

$$= \boxed{\frac{\sin a \cdot \cos \Delta x + \sin \Delta x \cdot \cos a - \sin a}{\Delta x}}$$

$$= \sin a \boxed{\frac{\cos \Delta x - 1}{\Delta x}} + \cos a \boxed{\frac{\sin \Delta x}{\Delta x}}$$

$$= \cos a,$$

since $\boxed{\sin \Delta x / \Delta x} = 1$ from p. 41, and $(\cos \Delta x - 1)/\Delta x$ is infinitesimal, as follows: if it were *not*, then since $(\cos \Delta x + 1)/\Delta x$ is infinite, the product

$$\frac{\cos \Delta x - 1}{\Delta x} \cdot \frac{\cos \Delta x + 1}{\Delta x} = \frac{\cos^2 \Delta x - 1}{(\Delta x)^2}$$

would be infinite. But this is simply $-\sin^2 \Delta x/(\Delta x)^2 = -(\sin \Delta x/\Delta x)^2$, which is infinitely close to -1. Thus, $(\cos \Delta x - 1)/\Delta x$ must be infinitesimal.

EXERCISES
Find the derivatives:
1. cos x.
2. tan x.
3. sec x.
4. cot x.
5. csc x.

For an example of the Chain Rule at work, we find the derivative of sin x^2: In this case, $f(x)$ is x^2 and $g(y)$ is sin y, so

sin $x^2 = g(f(x))$.

Applying the theorem,

$[\sin x^2]' = g'(f(x))f'(x)$

$$= (\cos x^2)\, 2x.$$

EXERCISES
Use theorem 7.5 to differentiate:
1. $(x + 3)^2$.
2. $\sqrt{x^2 + 7}$.
3. $1/\sqrt{x}$.
4. Once again, assuming $f(x) = \sqrt{x}$ has a derivative $f'(x)$, prove that it is $1/(2\sqrt{x})$ by letting $g(x) = x^2$ and using theorem 7.5.

PROOF: Let $q(x)$ be the function $1/g(x)$. Then $h(x) = f(x)q(x)$, so

$h'(b) = f'(b)q(b) + f(b)q'(q)$

$$= \frac{f'(b)}{g(b)} + f(b)\left(- \frac{g'(b)}{[g(b)]^2}\right)$$

$$= \frac{f'(b)g(b) - f(b)g'(b)}{[g(b)]^2}. \qquad \square$$

THEOREM 7.5 (the Chain Rule). If f is differentiable at b and g is differentiable at $f(b)$, then the function $h(x) = g(f(x))$ is differentiable at b and its derivative is $g'(f(b)) \cdot f'(b)$.

PROOF: Let Δx be any infinitesimal, and let

$\Delta f = f(b + \Delta x) - f(b),$

$\Delta g = g(f(b) + \Delta f) - g(f(b)),$

and

$\Delta h = h(b + \Delta x) - h(b).$

Note that

$\Delta h = g(f(b + \Delta x)) - g(f(b)) = \Delta g.$

By theorem 7.1, f is continuous, so that Δf is either infinitesimal or zero. In either case we will show that

$$\boxed{\frac{\Delta h}{\Delta x}} = g'(f(b))f'(b).$$

(1) If Δf is zero, then so is Δh, so

$$\boxed{\frac{\Delta h}{\Delta x}} = 0 = g'(f(b)) \cdot 0 = g'(f(b))\boxed{\frac{\Delta f}{\Delta x}}$$

$$= g'(f(b))f'(b).$$

(2) If Δf is infinitesimal, then since $\Delta h = \Delta g$, we have

$$\boxed{\frac{\Delta h}{\Delta x}} = \boxed{\frac{\Delta g}{\Delta x}} = \boxed{\frac{\Delta g}{\Delta f}} \cdot \boxed{\frac{\Delta f}{\Delta x}}$$

$$= g'(f(b))f'(b). \quad \square$$

The next theorem, although more in the category of a specific example, is singled out as extremely important because of the key role polynomials play in mathematics.

5. Prove that $f(x) = |x|$ is continuous at 0. (*Hint*: You only have to show that the absolute value of an infinitesimal is infinitesimal.)

6. Prove that $f(x) = |x|$ is not differentiable at 0. (*Hint*: If $\Delta x < 0$, $\Delta f/\Delta x = -1$, and if $\Delta x > 0$, $\Delta x/\Delta f = +1$, so $\boxed{\Delta f/\Delta x}$ is not constant for all infinitesimal Δx.) (This is actually the full proof with only a few details missing.)

The discussion on p. 66 tells us that the derivative of h is 1.

This proof is actually a disguised proof by induction. We note that the theorem is true for $n = 1$. Then show that if the theorem is true for $n = k - 1$, it must be true for $n = k$.

Another variation of this is Fermat's method of "infinite descent." His approach was to show that if the theorem failed for some number k, it also failed for a smaller number (in this case, $k - 1$), and hence an even smaller number, and so on. Since there are no infinite descending sequences of natural numbers, we have a contradiction.

Another proof of theorem 7.6 is: For any Δx, let $h = a + \Delta x$ so that

$$\frac{\Delta f}{\Delta x} = \frac{h^n - a^n}{h - a}$$

$$= h^{n-1} + h^{n-2}a + h^{n-3}a^2 + \cdots$$
$$+ a^{n-1}.$$

If Δx is infinitesimal, then

$$\left|\frac{\Delta f}{\Delta x}\right| = \boxed{h^{n-1}} + \boxed{h^{n-2}\,a} + \cdots + \boxed{a^{n-1}}$$

$$= a^{n-1} + a^{n-1} + \cdots + a^{n-1}$$

$$= na^{n-1}.$$

EXERCISE

Use theorems 7.4 and 7.6 to prove the derivative of x^{-n} is $-nx^{-n-1}$ for $n > 0$. (*Hint*: Use $x^{-n} = 1/x^n$.)

THEOREM 7.6. For each positive integer n, the function x^n has derivative nx^{n-1}.

PROOF: Suppose the theorem is not true. Then for some n, x^n does not have derivative nx^{n-1}. Let k be the least such n. Note that $k > 1$ since the discussion on p. 66 proves the theorem for $k = 1$ or 0. Then as $k - 1 < k$, the theorem must be true for x^{k-1}, and so the derivative of x^{k-1} is $(k - 1)x^{k-2}$. We can use this fact, for if $g(x)$ is x^{k-1} and $h(x)$ is x, theorem 7.3 tells us that the derivative of $g(x)h(x)$ is

$$g'(x)h(x) + g(x)h'(x)$$

$$= (k - 1)x^{k-2}\,x + x^{k-1} \cdot 1$$

$$= kx^{k-1}.$$

But $g(x)h(x) = x^k$, and so we have shown that x^k has derivative kx^{k-1}, contradicting our choice of k. (Recall that k was defined to be the least number such that x^k *did not* have derivative kx^{k-1}.) $\square$

We conclude this chapter with two theorems of great importance. The first is the basis for the solution of an enormous category of physical problems dealing with maxima and minima. These problems provided much of the original motivation for the calculus. Similar problems motivated other branches of analysis and continue to do so even today.

The second theorem, the Mean Value Theorem, is the taking off point for dozens of theorems. Actually a corollary of the first theorem, it numbers among its uses:

1. Taylor's Theorem—essentially a generalization of the Mean Value Theorem—demonstrates how a large class of functions can be approximated infinitely well by polynomials;

2. A different generalization by Cauchy that establishes L'Hôpital's Rule;

3. Yet another generalization, this one providing a neat definition of the nth derivative of a function for any positive integer n;

4. The answer to the question: "If $f'(x)$ is constantly 0, is $f(x)$ constant?";

5. Half of the proof of the Fundamental Theorem of Calculus;

Darboux's Theorem states that if f is differentiable, then f' has the Darboux Property (see p. 47). As it happens, this is about as strong a result as possible, for, in general, f' may be neither differentiable nor continuous. One example of this is the function $f(x) = |x|x$:

6. Darboux's Theorem;

7. A tidy characterization of functions that have a continuous derivative.

Consider the following problem. The function $y = x^3 + 2x^2 + x + 1$ is continuous on the interval $[-\frac{3}{4}, -\frac{1}{4}]$.

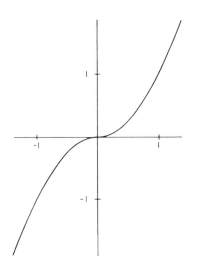

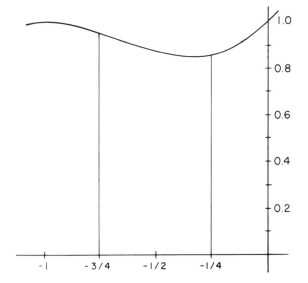

EXERCISES
(concerning the above function f):
1. Prove that for $x > 0, f'(x) = 2x$.
2. Prove that for $x < 0, f'(x) = -2x$.
3. Prove that $f'(0) = 0$.
4. Prove that $f'(x) = 2|x|$ is not differentiable at $x = 0$ (see p. 66).

An example of a function that is differentiable but whose derivative is not continuous is more complicated. (See p. 73.)

Given a function f, f' may, of course, be differentiable. In this case its derivative is called the second derivative of f and is denoted by f'' or $f^{(2)}$. In general, the nth derivative of f is written $f^{(n)}$.

Leibniz, when he eventually discovered how to differentiate $h(x) = f(x)g(x)$, also discovered a formula for its nth derivative:

$$h^{(n)}(x) = \sum_{i=0}^{n} \binom{n}{i} f^{(i)}(x) g^{(n-i)}(x).$$

Hence, by theorem 5.2, $y = x^3 + 2x^2 + x + 1$ assumes a minimum at some x in $[-\frac{3}{4}, -\frac{1}{4}]$. Find the x in $[-\frac{3}{4}, -\frac{1}{4}]$ at which the minimum is assumed. To one who knows no calculus this problem is extraordinarily difficult, but to us it is simple, for a differentiable function, at the point where it attains a minimum, should have a tangent with slope 0.

The derivative of $x^3 + 2x^2 + x + 1$ is $3x^2 + 4x + 1$, and this is 0 at

$$x = \frac{-4 \pm \sqrt{16 - 12}}{6} = -\frac{2}{3} \pm \frac{1}{3}.$$

Of these two points, only $x = -\frac{1}{3}$ is in the interval $[-\frac{3}{4}, -\frac{1}{4}]$, and this turns out to be the point at which the function assumes its minimum in that interval. (Consider, in contrast, the interval $[-1\frac{1}{4}, -\frac{3}{4}]$. At what x does the function assume its minimum here?)

This same technique was explored independently by Johannes Kepler, who was intrigued by the inaccurate but customary method for measuring the size of wine casks. He discovered that the citizens of his village, while ignorant of calculus, still managed to construct wine casks to take the greatest possible advantage of the measuring method.

Why is $c + \Delta x \in (a, b)$?

Assuming the function $f(x) = \log x$ is defined and has a derivative $f'(x)$, we can prove that $f'(x)$ is the function k/x for some constant k.

PROOF: Let $k = f'(1)$. Then for any real $b > 0$,

$$\log x + \log b = \log bx$$

and, differentiating, we obtain

$$f'(x) = bf'(bx)$$

(using theorem 7.5). Then letting $x = 1$, we get

$$f'(b) = \frac{f'(1)}{b} = \frac{k}{b}.$$

This technique, pioneered by Pierre de Fermat early in the seventeenth century, is crystallized in the following theorem:

THEOREM 7.7. Assume that $f(x)$ is differentiable on the open interval (a, b) and that at a point c in (a, b), f assumes a maximum (or a minimum). Then $f'(c) = 0$.

PROOF: Let us assume $f(c)$ is a maximum (a similar proof applies if it's a minimum). Since $f(c)$ is a maximum,

$$f(c + \Delta x) - f(c) \le 0$$

for every infinitesimal Δx. So let $(\Delta x)_1$ be a positive infinitesimal and $(\Delta x)_2$ a negative infinitesimal. Then

$$\frac{f(c + (\Delta x)_1) - f(c)}{(\Delta x)_1} \le 0 \le \frac{f(c + (\Delta x)_2) - f(c)}{(\Delta x)_2}.$$

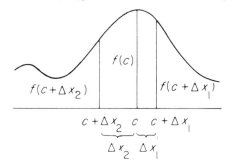

Thus

$$f'(c) = \left| \frac{f(c + (\Delta x)_1) - f(c)}{(\Delta x)_1} \right|$$

$$= \left| \frac{f(c + (\Delta x)_2) - f(c)}{(\Delta x)_2} \right|$$

$$= 0. \quad \square$$

This theorem is also called Lagrange's Theorem, after the French mathematician Joseph Louis Lagrange (1736–1813).

EXERCISE
Consider the function

$$f(x) = \begin{cases} x^2 & \text{if } x \text{ is rational} \\ -x^2 & \text{if } x \text{ is irrational.} \end{cases}$$

1. Prove that $f(x)$ is not continuous for $x \ne 0$.
2. Prove that $f(x)$ *is* differentiable at $x = 0$!

THEOREM 7.8 (the Mean Value Theorem). Assume that $f(x)$ is differentiable on the interval (a, b). Then for some c in (a, b),

$$\frac{f(b) - f(a)}{b - a} = f'(c).$$

PROOF: This is just a rotated version of the previous theorem.

The simplest example of a function that is differentiable but whose derivative is discontinuous resembles certain functions introduced in Chapter 5. Let f be defined by

$$f(x) = \begin{cases} x^2 \sin \dfrac{1}{x} & \text{if } x \neq 0 \\ 0 & \text{if } x = 0. \end{cases}$$

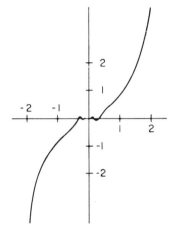

First, f is differentiable at all points $x \neq 0$ by theorems 7.3, 7.4, and 7.5. At $x = 0$,

$$\left| \frac{f(0 + \Delta x) - f(0)}{\Delta x} \right| = \left| \Delta x \sin \frac{1}{\Delta x} \right|$$

$$\leq |\Delta x| \approx 0,$$

since $-1 \leq \sin(1/\Delta x) \leq 1$. Thus

$$\frac{f(0 + \Delta x) - f(0)}{\Delta x} = 0$$

for all infinitesimal Δx, and hence f is differentiable. The derivative is given by

$$f'(x) = \begin{cases} 2x \sin \dfrac{1}{x} - \cos \dfrac{1}{x} & \text{if } x \neq 0 \\ 0 & \text{if } x = 0 \end{cases}$$

(by theorems 7.3, 7.4, 7.5).

EXERCISE
Prove that the function f' given above is discontinuous at 0. (*Hint*: See p. 46.)

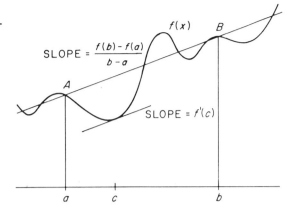

We note that

$$\frac{f(b) - f(a)}{b - a}$$

is just the slope of the chord $\overline{AB}$. If we rotate the function f so that $\overline{AB}$ becomes part of the x-axis,

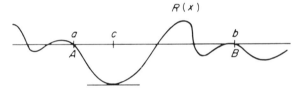

we merely must find a c such that

$$R'(c) = 0$$

for the rotated function $R(x)$. But by theorem 7.7, this c will be a point in (a, b) where R assumes the maximum or minimum, a point we know to exist by a previous theorem concerning continuous functions.

Now how can we make all of this precise and prove our theorem? Given our function $f(x)$, let $R(x)$ be defined by

$$R(x) = f(x) - f(a) - \frac{f(b) - f(a)}{b - a}(x - a).$$

Note that

$$R'(x) = f'(x) - \frac{f(b) - f(a)}{b - a}$$

using theorem 7.2 and the example on p. 66. Thus $R(x)$ is our desired rotated version of f. Let c be a point at which R assumes its maximum (or mini-

One of the clearest and most valuable applications of derivatives involves speed and distance. Suppose a car drives along a straight road and its distance from a fixed point at any time t is given by $s(t)$. Then at any time t_0, what is its velocity? The problem is that the velocity may be changing. We can't simply take another point in time t_1 and compute

$$\frac{s(t_1) - s(t_0)}{t_1 - t_0},$$

for this will only give us the car's average velocity over the time interval $[t_0, t_1]$. On the other hand, if we let t_1 be *infinitely close* to t_0, then the average velocity on $[t_0, t_1]$ will be *infinitely close* to the velocity of the car at t_0. Thus the speed of the car is

$$\approx \frac{s(t_0 + \textcircled{\circ}) - s(t_0)}{\textcircled{\circ}}$$

or $s'(t_0)$.

This use of the differential calculus was first discovered by Galileo.

Note: We already know the converse of this theorem—if $f(x) = k$, a constant, then $f'(x) = 0$. If we interpret this theorem in terms of distance and velocity, we have proven mathematically in (1) that if a car's speed is constantly 0, the car won't go anywhere.

mum). If $a < c < b$, our previous theorem says that $R'(c) = 0$, that is,

$$f'(c) = \frac{f(b) - f(a)}{b - a}.$$

If, on the other hand, both the maximum and minimum of $R(x)$ occur at the end-points a and b, then since

$$R(a) = f(a) - f(a) - \frac{f(b) - f(a)}{b - a}(a - a) = 0$$

and

$$R(b) = f(b) - f(a) - \frac{f(b) - f(a)}{b - a}(b - a) = 0,$$

the maximum value of R equals the minimum value of R (on $[a, b]$), and so R must be constant (on $[a, b]$).

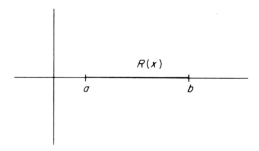

Thus $R'(c) = 0$ for *any* c in $[a, b]$, and again,

$$f'(c) = \frac{f(b) - f(a)}{b - a}. \quad \square$$

Here is a quick application:

COROLLARY. Assume that f is differentiable on $[a, b]$. Then
(1) if $f'(x) = 0$ on $[a, b]$, f is constant.
(2) if $f'(x) > 0$ on $[a, b]$, f is monotone, nondecreasing, and is *never* constant.
(3) if $f'(x) < 0$ on $[a, b]$, then f is monotone, nonincreasing, and is *never* constant.

PROOF: Let $c < d$ be numbers in $[a, b]$. By the Mean Value Theorem, there is a q between c and d, such that

$$f'(q) = \frac{f(d) - f(c)}{d - c}$$

and so $(d - c)f'(q) = f(d) - f(c)$. If $f'(x)$ is always 0, then $f(d) - f(c) = 0$. So $f(d) = f(c)$. This proves (1).

If $f'(x)$ is always positive, then so is $f(d) - f(c)$. Hence $f(d) > f(c)$. This proves (2).

Finally, if $f'(x)$ is always negative, $f(d) - f(c)$ is negative, and so $f(d) < f(c)$, proving (3). □

This corollary has an important corollary:

COROLLARY (the Inverse Function Theorem). Suppose that f is differentiable, and either $f'(x) > 0$ on $[a, b]$ or $f'(x) < 0$ on $[a, b]$. Then f has a continuous inverse function g, defined between $f(a)$ and $f(b)$, and for all $y = f(x)$,

$$g'(y) = \frac{1}{f'(x)}.$$

[Recall from analytic geometry that g is an inverse function for f if $f(g(y)) = y$ and $g(f(x)) = x$ for all x and y in the appropriate domains.]

PROOF: By theorem 7.1, f is continuous. Thus, for any y between $f(a)$ and $f(b)$, there is an x between a and b such that $f(x) = y$ by the Intermediate Value Theorem. Since f is never constant by the previous corollary, there is only one such x. Let us define $g(y)$ to be that x. With this definition we have

$$f(g(y)) = y$$

and

$$g(f(x)) = x,$$

and so g is the inverse of f.

To see that g is continuous, suppose y is a given real, and $h \approx y$. Then if $g(h) \not\approx g(y)$, there must be a real number r between $g(y) = x$ and $g(h) = q$. Then $f(r)$ must be between $y = f(g(y))$ and $h = f(g(h))$, since f is monotone. But since $y \approx h$, there are no reals between y and h. Thus there can be no real between $g(y)$ and $g(h)$, and so $g(y) \approx g(h)$. Thus g is continuous.

Finally, to find $g'(y)$, let $\varDelta y$ be any infinitesimal, let $x = g(y)$ (hence $y = f(x)$), and let $\varDelta x = g(y + \varDelta y) - g(y)$. As we have just seen, $\varDelta x$ is infinitesimal. Thus

$$\frac{g(y + \varDelta y) - g(y)}{\varDelta y} = \frac{\varDelta x}{(y + \varDelta y) - y}$$

$$= \frac{\varDelta x}{f(g(y + \varDelta y)) - f(x)}$$

$$= \frac{\varDelta x}{f(x + \varDelta x) - f(x)}$$

$$= \frac{1}{f'(x)}. \quad □$$

Graphically, the inverse curve of a given curve

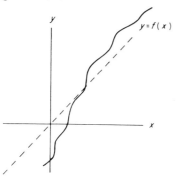

is given by flipping the curve over the line $y = x$ like this:

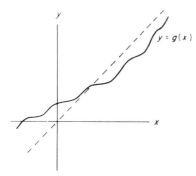

Geometrically, it is obvious that if f is continuous, so is g.

EXERCISES
Verify that the following functions are inverses:
1. x^2 and $\sqrt{y}$.
2. $2x - 5$ and $y/2 + 5/2$.
3. 10^x and $\log_{10} y$.
4. x^n and $y^{1/n}$.
Find the inverses:
5. $x + 7$.
6. $3x - 1$.
7. x^3.
8. $x\sqrt{x}$.
9. Use (4) to prove that the derivative of $g(y) = y^{1/n}$ is $g'(y) = (1/n)y^{1/n-1}$.
10. Use (9) and theorems 7.5 and 7.6 to prove that the derivative of $x^{n/m}$ is $(n/m)x^{n/m-1}$. (*Hint*: $h(x) = x^{n/m}$ $= g(f(x))$ where $f(x) = x^n$ and $g(y)$ $= y^{1/m}$.)

Note: On the interval $(0, \pi)$, the derivative of $f(x) = \cos x$ is negative, so $\cos x$ has an inverse, usually written $g(y) = \arccos y$, or $\cos^{-1} y$. By the inverse function theorem, we can find

$$g'(y) = \frac{1}{f'(x)} = \frac{1}{-\sin x}$$

$$= \frac{-1}{\sqrt{1 - \cos^2 x}}$$

$$= \frac{-1}{\sqrt{1 - y^2}}.$$

11. Find an interval where the inverse of $\sin x$ exists and find its derivative.
12. Find an interval where the inverse of $\tan x$ exists and find its derivative.

8

The Fundamental Theorem

Those impostors, then, whom they call
mathematicians, I consulted without
scruple, because they seemed to use no
sacrifice, nor pray to any spirit for
their divinations.
St. Augustine (354–430)

Thus far our work in the calculus has centered on two main notions, integration and differentiation. These two operations that one can perform on functions seem to bear no connection with one another, and indeed they were studied as independent concepts for almost 2,000 years. How amazing, then, that it should be discovered in the late seventeenth century that they are in fact intimately related. Loosely speaking, it turned out that if one took a function, integrated it, and differentiated the result, one got the original function back. Similarly, if one takes a function and differentiates it, the integral of this is again the original function. The importance of this can hardly be understated. Not only is this an elegant unification of two branches of mathematics, but it replaces the very difficult process of integration with the relatively easy process of differentiation.

Here is the theorem:

THE FUNDAMENTAL THEOREM OF CALCULUS. Let f be any continuous function. Then for any function H,

H is a primitive of f

iff

f is the derivative of H.

Before proving the theorem, let us take an example illustrating its enormous practical power:

Suppose we wish to evaluate

$$\int_{-5}^{2} (7x^6 + 12x^5 + 20x^3 + 7)\, dx.$$

Using the machinery from Chapter 6, this would be a virtually impossible task. However, with our new theorem, it is almost trivial. By inspection the derivative of

Recall that the definition of F being a primitive of f is "for any a, b,

$$\int_{a}^{b} f(x)\, dx = F(b) - F(a).\text{''}$$

The Fundamental Theorem then tells us that for any differentiable function g and any a, b,

$$\int_{a}^{b} g'(x)\,dx = g(b) - g(a).$$

The Fundamental Theorem of Calculus is the "one-dimensional" version of a fact known as Stokes's Theorem (the two-dimensional version of Stokes's Theorem is often referred to as Green's Theorem). Loosely speaking, Stokes's result says that given a region in n-dimensional space (for any finite n), one can evaluate the integral of a derivative of a function throughout the interior of the region simply by evaluating the integral of the function itself on the boundary of the region. In one dimension the interval $[a, b]$ is a typical region, and the two points a and b form its boundary.

The wonder of this theorem is that it connects two entirely different geometric ideas—tangents and areas. It is perhaps the unexpectedness of this that prevented its discovery for so long. As long as the problem was viewed *geometrically*, the secret of the calculus remained obscure. The key is to view it *algebraically*, in terms of functions. As the eighteenth-century mathematician Lagrange put it:

So long as algebra and geometry proceeded separately their progress was slow and their application limited, but when these two sciences were united, they mutually strengthened each other, and marched together at a rapid pace toward perfection.

Not everyone saw it this way, however. Thomas Hobbes (1588–1679) (see note, p. 13) liked his geometry pure and objected to "the whole herd of them who apply their algebra to geometry." This in fact was the basis of his pique at Wallis's "scurvy book."

In physics a well-known law credited to Gauss tells us that to determine the amount of charge contained within a given region of space, it suffices merely to integrate the electromagnetic force field around the boundary of the region. This law, the Fundamental Theorem, and, most generally, Stokes's Theorem, are all versions of the same phenomenon.

$$x^7 + 2x^6 + 5x^4 + 7x$$

is

$$7x^6 + 12x^5 + 20x^3 + 7,$$

and so by the Fundamental Theorem,

$$x^7 + 2x^6 + 5x^4 + 7x$$

is a primitive of

$$7x^6 + 12x^5 + 20x^3 + 7.$$

By the definition of primitive, then,

$$\int_{-5}^{2} (7x^6 + 12x^5 + 20x^3 + 7) \, dx$$

is equal to

$$[(2)^7 + 2(2)^6 + 5(2)^4 + 7(2)]$$
$$- [(-5)^7 + 2(-5)^6 + 5(-5)^4 + 7(-5)]$$
$$= 350 - (-43785) = 44135.$$

And now the proof:

PROOF: Assume H is a primitive of f.

Let a be any real number and assume $\Delta x > 0$, Δx real. Since f is continuous, f assumes a maximum M and minimum m on $[a, a + \Delta x]$.

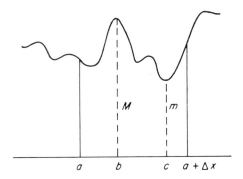

Suppose $f(b) = M$ and $f(c) = m$. Then we have

$$m \, \Delta x \le \int_{a}^{a+\Delta x} f(t) dt \le M \, \Delta x,$$

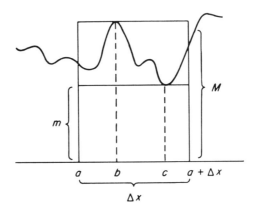

that is,

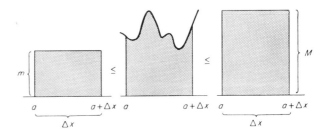

The quotation at the head of this chapter is taken from the *Confessions* of St. Augustine. After his reform and conversion to Christianity, St. Augustine had little good to say of his youthful pursuits and companions. He was particularly rough on mathematicians:

The good Christian should beware of mathematicians and all those who make empty prophecies. The danger already exists that the mathematicians have made a covenant with the Devil to darken the spirit and to confine man in the bonds of Hell . . .

The Fundamental Theorem was discovered independently by a number of mathematicians who by and large did not realize its true importance. Gregory published it in geometric form in 1668. Isaac Barrow, Newton's teacher, proved it for monotonic functions at about the same time. Torricelli (1608–1647)

Since H is a primitive of f, this becomes

$$m \, \Delta x \leq H(a + \Delta x) - H(a) \leq M \, \Delta x$$

or

$$f(c) \leq \frac{H(a + \Delta x) - H(a)}{\Delta x} \leq f(b).$$

We have just shown that "for all $\Delta x < 0$ there exist b and c in the interval $[a, a + \Delta x]$ such that

$$f(c) \leq \frac{H(a + \Delta x) - H(a)}{\Delta x} \leq f(b)."$$

Since this sentence is true in **R** and can be written in our language L, it must also be true in **HR**. Thus, for any *infinitesimal* $\Delta x > 0$, there exist b, c in $[a, a + \Delta x]$ such that

$$f(c) \leq \frac{H(a + \Delta x) - H(a)}{\Delta x} \leq f(b).$$

probably understood the theorem also, but his work wasn't actually published until 1918. It is thought that even Galileo (1564–1642) had some understanding of it. These glimpses of truth were finally assembled by Newton and Leibniz.

Barrow, known as a wit and an athlete at Cambridge, realized Newton's ability early. In recognition of the brilliance of his student, he resigned his professorship in favor of Newton and never worked again in mathematics.

The first *rigorous* proof of this theorem is due to Augustin Cauchy (1789–1857).

(See theorem 6.4.)

. . . rigor is the concern of philosophy, and not of geometry.
Francesco Cavalieri (1598–1647)

The brothers Johann and Jakob Bernoulli did much to explain and expand the methods of Leibniz. In fact, Leibniz himself said the calculus was as much theirs as his.

But since $a \approx a + \Delta x$, we have

$$a \qquad b \qquad c \qquad a + \Delta x$$

and $b \approx a \approx c$, so by the continuity of f,

$$f(b) \approx f(a) \approx f(c)$$

and so

$$\frac{H(a + \Delta x) - H(a)}{\Delta x} \approx f(a).$$

As $f(a)$ is a real number, we must have

$$\left| \frac{H(a + \Delta x) - H(a)}{\Delta x} \right| = f(a).$$

Similarly, if $\Delta x < 0$, we can derive the same equation, hence

$$H'(a) = f(a).$$

Now to prove the converse: Suppose $H(x)$ is a function such that $H'(x) = f(x)$. Let $G(x)$ be the function

$$\int_0^x f(t)\, dt.$$

Then $G(x)$ is a primitive of f and so by the first half of our proof

$$G'(x) = f(x).$$

As both H and G have derivative f, $H - G$ must have derivative 0 (theorem 7.2), and so by theorem 7.9, $H - G$ is a constant. Let k be the constant. Then since

$$H(x) = G(k) + k,$$

theorem 6.4 tells us that H is a primitive of f. $\quad\square$

One outstanding difference between integral and differential calculus which you may have noticed in Chapters 6 and 7 is that one is fairly difficult and the other is more manageable. As a result we have many more theorems about differentiation than integration. The Fundamental Theorem can do much to even things up. Guided by it, every theorem on one side can be connected to a theorem on the other. For example, from the result $(f + g)' = f' + g'$ (theorem 7.2) we

This theorem may also be proved directly from Chapter 6 by noting that for any Δx,

$$\overset{b}{\underset{a}{S}}\,[f(x) + g(x)]\,\Delta x$$

$$= \overset{b}{\underset{a}{S}}\,f(x)\Delta x + \overset{b}{\underset{a}{S}}\,g(x)\Delta x.$$

(See p. 55.)

may deduce

$$\int_a^b [f(x) + g(x)]dx = \int_a^b f(x)dx + \int_a^b g(x)dx.$$

PROOF: Let F and G be primitives for f and g, respectively. Translating the above theorem with the Fundamental Theorem, we get that the sum of primitives for f and g ($F + G$) is a primitive of $f + g$. Thus,

$$\int_a^b [f(x) + g(x)]dx = F(b) + G(b) - F(a) - G(a)$$

$$= F(b) - F(a) + G(b) - G(a)$$

$$= \int_a^b f(x)dx + \int_a^b g(x)dx. \qquad \square$$

In regard to the correspondence given us by the Fundamental Theorem, it might be useful to make a list:

Differential Calculus

THEOREM 1. If f is differentiable, then f is continuous.

THEOREM 2. $(f + g)' = f' + g'$.

THEOREM 3. $(fg)' = f'g + fg'$.

THEOREM 4. $(1/f)' = f'/f^2$.

THEOREM 5. $g(f)' = g'(f)f'$.

THEOREM 6. $(x^n)' = nx^{n-1}$.

THEOREM 7. The derivative of f is 0 at any maximum or minimum point in an open interval.

THEOREM 8. The Mean Value Theorem.

THEOREM 9. If $f'(x) = 0$, then f is a constant function.

THEOREM 10. If $d(t)$ represents the distance of an object from a fixed point at time t, then $d'(t)$ represents the velocity of that object at time t.

Integral Calculus*

$$\int_a^b [f(x) + g(x)]dx = \int_a^b f(x)dx + \int_a^b g(x)dx.$$

Integration by parts.

Integration by substitution.

EXERCISE
Fill in the blanks in the chart.

The theorem corresponding to (6) was known in varying degrees by Fermat, Cavalieri, Roberval, Torricelli, and Wallis.

Integration by parts and integration by substitution are two very important techniques with which the reader should be familiar. It is a good exercise to prove their validity using the Fundamental Theorem and the corresponding differential calculus theorems.

*The appropriate integration theorems missing in the list may be supplied by the reader. They vary in interest and usefulness from very valuable (6) to practically useless (9).

The Fundamental Theorem

The theorem corresponding to (8) is often called the Mean Value Theorem for Integrals.

Logarithms were first invented by Napier in 1614 as a computational aid.

Napier also invented a mechanical device to help in multiplying integers. The device consisted of long sticks with numbers on them called "Napier's bones."

The first to notice a connection between $1/x$ and logs was Gregory de St. Vincent, who found

$$\int_{x_0}^{x} \frac{dt}{t} = k \log \frac{1}{x}.$$

EXERCISE

Can this be right? What is k? (Use 7.)

That log is the integral of $1/x$ was discovered by his student, Alfons A. de Sarasa.

If a bank offers depositors "6% interest, compounded annually" this means that given a deposit of $347, it will return to you, after a year, $347 plus 6%, that is,

$(347)(1.06) = \$367.82.$

If the bank offers the same interest rate *compounded twice a year*, this means that it will give you 3% interest *twice a year*, that is, after 6 months, you will have

$(347)(1.03) = \$357.41,$

and at the end of the year,

$(357.41)(1.03) = \$368.13.$

Similarly, if the interest is compounded 3 times a year, you will have

$(347)(1.02)(1.02)(1.02) = \$368.24$
(pocket calculator),

6 times a year:

$(347)\,(1.01)^6 = \$368.35,$

12 times a year:

$(347)\,(1.005)^{12} = \$368.40.$

Notice that the money you receive from these various plans increases

The following problems use calculus to develop the log and exponential functions.

EXERCISES

1. Prove that $1/x$ is continuous for all $x > 0$. Conclude that $1/x$ is integrable on all positive intervals (theorem 6.2).

DEFINITION. For $x > 0$, let

$$\ln x = \int_{1}^{x} \frac{1}{t}\, dt.$$

$\ln x$ is called "the natural logarithm of x" and is sometimes written $\log x$.

2. Prove that $\ln x$ is a primitive of $1/x$ for positive x and note that $\ln 1 = 0$ (see theorem 6.4).

3. Prove that the derivative of $\ln x$ is $1/x$ for all $x > 0$.

4. Use the Chain Rule to prove that if k is any constant, the derivative of $\ln kx$ is $1/x$.

5. Use the corollary, p. 74, to prove that if k is any constant, there is a constant C such that

$$\ln kx = C + \ln x.$$

(*Hint*: Let $f(x) = \ln kx - \ln x$ and find $f'(x)$ using exer. 5, p. 67.)

6. Prove that $\ln kx = \ln k + \ln x$. (*Hint*: Use 5 and let $x = 1$, note that $\ln 1 = 0$.) Conclude that for all positive a, b,

$$\ln ab = \ln a + \ln b.$$

7. Use 6 to prove that for all positive a and b, $\ln (a/b) = \ln a - \ln b$. (Let $x = a/b, k = b$.)

8. If r is any rational number, prove that the derivative of the function $\ln x^r$ is r/x. Use the Chain Rule, and exer. 10, p. 75.

9. Prove that if r is any rational number, there is a constant c such that $\ln x^r = c + r \ln x$.

10. Prove that for rational r, $\ln x^r = r \ln x$.

11. Use the Inverse Function Theorem to prove that $\ln x$ has an inverse function.

DEFINITION. Let exp denote the inverse function of $\ln$, so that for all appropriate numbers a, $\exp(\ln a) = a$ and $\ln(\exp a) = a$.

12. Let $b = \exp a$, $d = \exp c$ (so that $\ln b = a$ and $\ln d = c$). Use 6 to prove: $\exp(a + c) = \exp a \cdot \exp c$.

with the frequency of compounding, though the increases are getting smaller.

Finally, some banks offer "continuous compounding." This can be expressed with hyperreals as "compounded N times a year," where $N > 0$ is infinite:

$$A = (347)\left(1 + \frac{.06}{N}\right)^N.$$

To calculate A we take the log of it:

$$\ln A = \ln 347 + N\ln\left(1 + \frac{.06}{N}\right)$$

(by 6 and 10)

$$= \ln 347$$
$$+ N\left[\ln\left(1 + \frac{.06}{N}\right) - \ln 1\right]$$

$$= \ln 347$$
$$+ \frac{.06\left[\ln\left(1 + \frac{.06}{N}\right) - \ln 1\right]}{.06/N}.$$

Now let $\Delta x = .06/N$, an infinitesimal, and we have

$$= \ln 347$$
$$+ .06\left[\frac{\ln 1 + \Delta x - \ln 1}{\Delta x}\right].$$

This last is infinitely close to the derivative of $\ln x$ at 1, which is $1/1$ or 1. Thus

$$\ln A \approx \ln 347 + .06.$$

Taking exp of both sides,

$$\exp(\ln A) \approx \exp(\ln 347 + .06),$$

$$A \approx 347 \exp .06,$$

$$A \approx 347\, e^{.06}.$$

Since banks only pay out real amounts of money, you will receive

$$\lfloor A \rfloor = 347\, e^{.06}$$

$$= \$368.46.$$

($e^{.06}$ is approximately 1.061837; see p. 99.)

13. Prove $\exp(a - c) = \exp a/\exp c$.
14. Let $b = \exp a$, r rational. Prove that $\exp ar = (\exp a)^r$.

DEFINITION. Let $e = \exp 1$ (so $\ln e = 1$).
15. Prove that for all $a > 0$, r rational, $a^r = \exp(r \ln a)$.

DEFINITION. For *all* real numbers r, and $a > 0$, let $a^r = \exp(r \ln x)$.

16. Prove that $e^x = \exp x$.
17. Prove that $a^b \cdot a^c = a^{b+c}$. (Use 12.)
18. Prove that $a^b/a^c = a^{b-c}$.
19. Prove that $(a^b)^c = a^{bc}$.
20. Prove that the derivative of the function x^r is rx^{r-1}, for all reals r.
21. Notice that $\ln x$ is what in high school is called $\log_e x$, that is, the inverse function of e^x.

9

Infinite Sequences and Series

Despite the fact that the basic concepts of differentiation and integration are now quite natural and easy to understand, and despite the fact that the Fundamental Theorem tells us that integrating functions is no more difficult than finding primitives, it remains true that all but the simplest functions are quite difficult to work with. For example, it is easy to see that the function

$$5x^2 - 2x + 1 \tag{1}$$

has as derivative

$$10x - 2,$$

and

$$x^3 - 6x^2 + 2x + 7 \tag{2}$$

has as primitive

$$\frac{x^4}{4} - 2x^3 + x^2 + 7x,$$

but it is not at all easy to find the derivative of

$$\sqrt{\frac{1}{x - \sin \sqrt{x}}} \tag{3}$$

or the primitive of

$$\frac{1}{x^5 + 1}. \tag{4}$$

The feature that makes (1) and (2) easy to deal with is that they are polynomials. One can, in fact, state general rules for finding the derivatives and primitives of polynomials:

RULE. If $p(x) = a_n x^n + a_{n-1} x^{n-1} + \cdots + a_1 x + a_0$ is any given polynomial, then the derivative of $p(x)$ is

$$na_n x^{n-1} + (n - 1)a_{n-1} x^{n-2} + \cdots + a_1$$

and a primitive of $p(x)$ is

$$\frac{a_n}{n + 1} x^{n+1} + \frac{a_{n-1}}{n} x^n + \cdots + \frac{a_1}{2} x^2 + a_0 x.$$

This is hardly a rule to memorize, but it does exist and

similar rules for classes of functions other than polynomials simply do not. Wouldn't it be nice then if every function $f(x)$ were actually equal to some polynomial?

For a start, we shouldn't hope for literally *every* function to be equal to a polynomial, for polynomials are differentiable whereas there exist many functions that are not. There is an additional obstacle. Given any polynomial $p(x)$, if we differentiate it to get $p'(x)$, then differentiate $p'(x)$ to get $p''(x)$, and continue in this way, we eventually get a function that is identically equal to 0. Thus, for example, the function $y = \sin(x)$ can never be equal to a polynomial since no matter how often we differentiate $\sin(x)$ it is easy to see we are never left with a function that is identically zero.

The problem here lies in the fact that by differentiating again and again, a sufficiently large number of times, we can always render $y = x^n$ identically 0 *so long as n is finite.* But what if we allow n to be infinite, that is, what if we look at infinite polynomials? This is our objective.

We begin by examining the simpler notion of infinite sequences of numbers.

Intuitively, sequences are things that look like:

1, 2, 3, 4, 5, ...,

1/2, 1/4, 1/8, 1/16, 1/32, ... ,

or

$a_1, a_2, a_3, a_4, a_5, \ldots,$

but to be mathematically precise, we can define them as follows:

We often write a sequence in terms of the function defining it. For example, the sequence

(a)   1, 1/2, 1/3, 1/4, 1/5, ...

is denoted by

$\left\{ \dfrac{1}{n} \right\},$

the sequence

(b)   1/2, 1/4, 1/8, 1/16, ...

is denoted by

$\left\{ \dfrac{1}{2^n} \right\},$

DEFINITION. An *infinite sequence* is a function from the set of positive integers into the set of reals.

To correlate this "precise" definition with our intuitive concept, we think of any such function f as corresponding to the sequence

$f(1), f(2), f(3), f(4), f(5), \ldots.$

We will use $\{a_n\}$ to denote the sequence $a_1, a_2, a_3, a_4, a_5, \ldots$ Note that for $N > 0$ infinite, the "Nth term" of the sequence, a_N, is just $f(N)$, a number that exists since f appears in our language.

and the sequence

(c)   4, 9/2, 16/3, 25/4, ...

is denoted by

$$\left\{ \frac{(n + 1)^2}{n} \right\}.$$

The sequences given here all follow nice patterns, but it is well to keep in mind that sequences don't *have* to follow any pattern at all. The numbers in a sequence may be completely random.

Considering the three sequences just mentioned in terms of these two definitions,

(a) $1/n \to 0$, for if $N > 0$ is infinite, the Nth term in the sequence, $1/N$, is infinitely close to 0. $\{1/n\}$ is also bounded, since $1/N$ is finite.

(b) $1/2^n \to 0$, for if $n > 0$ is any integer, $1/2^n < 1/n$, and so $1/2^N < 1/N$ is infinitesimal. Similarly $\{1/2^n\}$ is bounded.

(c) $\{(n + 1)^2/n\}$ is not bounded, for the Nth term of the sequence,

$$\frac{(N + 1)^2}{N} = \frac{N^2 + 2N + 1}{N}$$

$$= N + 2 + \frac{1}{N},$$

is not finite. Clearly, it also fails to converge.

Not all bounded sequences converge. Consider the sequence

(d) 0, 1, 0, 1, 0, 1, 0, ... What is a_N, the Nth term of the sequence? If N is odd, a_N will be 0, and if N is even, a_N will be 1. Whichever it is, it is finite, so the sequence is bounded. On the other hand, there is no number such that $a \approx a_N$ for *all* infinite N. Thus the sequence does not converge.

Decide if the following sequences are bounded and/or convergent. If convergent, find the limit.

1. $\{n\}$.

2. $\left\{ \dfrac{1}{n^2} \right\}$.

3. $\left\{ \dfrac{1 + n}{n} \right\}$.

4. $\left\{ \dfrac{n}{n + 1} \right\}$.

5. $\{(1/3)^n\}$.

6. $\left\{ \dfrac{2n - 1}{4 - n} \right\}$.

DEFINITION. A sequence $\{a_n\}$ *converges* to a real a iff $a_N \approx a$ for all infinite integers $N > 0$. We will sometimes write this $a_n \to a$, or $\lim a_n = a$, and we say the *limit* of the sequence is a. If a sequence dosen't converge, we say it *diverges*.

DEFINITION. A sequence a_n is said to be *bounded* if a_N is finite for all infinite integers $N > 0$.

THEOREM 9.1. Any sequence converges to at most *one* real number, that is, if $\{a_n\}$ converges to a, and if it also converges to b, then $a = b$.

PROOF: Let $N > 0$ be an infinite integer. Then $a \approx a_N \approx b$. As $a \approx b$ and a and b are real, $a = b$. □

THEOREM 9.2. If $\{a_n\}$ is a sequence which converges, then $\{a_n\}$ is bounded.

PROOF: Suppose $\{a_n\}$ converges to a. Then $a_N \approx a$ for all infinite N and so, as a is real, a_N is finite for all infinite N. □

THEOREM 9.3. Suppose $\{a_n\}$ and $\{b_n\}$ are two sequences that converge to a and b, respectively. Then the sequences $\{a_n + b_n\}$ and $\{a_n \cdot b_n\}$ are convergent and converge to $a + b$ and $a \cdot b$, respectively. If, furthermore, $b \neq 0$, then $\{a_n/b_n\}$ converges to a/b.

PROOF: Exercise.

Our definition of what it means for a sequence to converge might be viewed by some as not sufficiently general in that it is given in terms of a real to which the sequence is converging. In other words, is it possible to give a definition of what it means for a sequence to converge based solely on the sequence itself with no mention of an additional real? In our context the answer is yes.

DEFINITION. A sequence $\{a_n\}$ converges iff for any infinite N and M, $a_N \approx a_M$.

This definition is due to Cauchy. To see that it is equivalent to our earlier definition we proceed as follows:

DEFINITION. A sequence $\{a_n\}$ is said to be a *Cauchy*

$$\frac{2N-1}{4-N} = \frac{\dfrac{2N}{N} - \dfrac{1}{N}}{\dfrac{4}{N} - \dfrac{N}{N}} = \frac{2 - \dfrac{1}{N}}{\dfrac{4}{N} - 1} \approx -2.$$

7. $\left\{\dfrac{2n+5}{3n+6}\right\}.$ 8. $\left\{\dfrac{n^2 - 7n + 6}{5n - 3n^2 + 2}\right\}.$

9. Prove theorem 9.3. (*Hint*: For example, the Nth term of the sequence $a_n + b_n$ is $a_N + b_N$, which is infinitely close to $a + b$ since $a_N \approx a$ and $b_N \approx b$.)

The sequence $\{n\}$ does not converge since the Nth term N is not infinitely close to any real number. This is different from the sequence $0, 1, 0, 1, 0, 1, \ldots$ where the Nth term is finite but not always the same. In the first case, we say the limit is ∞, sometimes described by saying " $\{n\}$ diverges to infinity."

DEFINITION. For a sequence $\{a_n\}$, we say $a_n \to \infty$ iff $a_N > 0$ is infinite for all infinite N. We say $a_n \to -\infty$ iff $a_N < 0$ is infinite for all infinite N.

EXERCISES
Find out if the following sequences have infinite limits.
1. $\{2^n\}$.
2. $\{\sqrt{n}\}$.
3. $1, \frac{1}{2}, 3, \frac{1}{4}, 5, \frac{1}{6}, \ldots.$
4. $1, -2, 3, -4, 5, -6, \ldots.$
5. Prove that if $a_n \to \infty$ or $a_n \to -\infty$, then $1/a_n \to 0$.
6. Find a sequence $\{a_n\}$ such that $a_n \to 0$ but $\{1/a_n\}$ has no finite or infinite limit.

sequence iff for any infinite N and M, $a_N \approx a_M$.

THEOREM 9.4. Every Cauchy sequence converges, that is, if $\{a_n\}$ is a Cauchy sequence, then there exists a real number a such that for every infinite N, $a_N \approx a$. (This property of the real numbers is known as *completeness*.) Furthermore, we conversely have that if $\{b_n\}$ is any sequence that converges to a real b, then $\{b_n\}$ is a Cauchy sequence.

PROOF: Suppose $\{a_n\}$ is a Cauchy sequence. We first note that for some real number $r > 0$,

(*) $-r < a_n < r$ for all n.

This is true, for if not, then the sentence

$\forall n \, \exists m [m > n \land |a_m| > |a_n| + 1]$

would be true in the reals and hence in the hyperreals. But for N infinite,

$|a_m| > |a_N| + 1$

is false for all $m > N$ (since $\{a_n\}$ is a Cauchy sequence).

Since (*) is true, a_N is finite for any infinite $N > 0$. Since $\{a_n\}$ is Cauchy, it is now immediate that

$\overline{a_N} \approx a_M$

for all infinite M and N. Thus $\{a_n\}$ converges to $\boxed{a_N}$.
The proof of the converse is left as an exercise. ☐

This next definition and theorem will be of great use to us later.

DEFINITION. A sequence $\{a_n\}$ is said to be *monotonic increasing* if $a_n \leq a_m$ whenever $n < m$

and is said to be *monotonic decreasing* if $a_n \geq a_m$ whenever $n < m$

A sequence is *monotonic* if it is either monotonic increasing or monotonic decreasing.

THEOREM 9.5. All monotonic bounded sequences converge.

Here is a corollary to this theorem: If $0 \le r < 1$, then $r^n \to 0$.

PROOF: Since $0 \le r^n < 1$ for all r, r^N is finite, so the sequence is bounded. The sequence is also monotonic decreasing, so by theorem 9.5, $r^n \to a$ for some real number a. If N is infinite, $r^N \approx a$, so $r^{N+1} = r^N \cdot r \approx ar$. But $N + 1$ is an infinite integer too, so $r^{N+1} \approx a$. Thus $ar \approx a$, and as ar and a are both real numbers, they must be equal. Since $r \ne 1$, we must have $a = 0$. $\square$

(For another proof, see p. 123.)

EXERCISES
1. Use the above to prove that $r^n \to 0$ for all r such that $|r| < 1$.
2. Prove that $\{r^n\}$ is monotonic decreasing whenever $0 \le r < 1$.

Sequences of the form $\{r^n\}$ are called *geometric sequences*, and, as we shall see later, they are particularly useful.

PROOF: Suppose $\{a_n\}$ is monotonic increasing and bounded (the proof is similar if $\{a_n\}$ is decreasing). We will show that $\{a_n\}$ is a Cauchy sequence. If $\{a_n\}$ is *not* Cauchy, then $a_N \napprox a_M$ for some infinite $N > M$. But since $a_N > a_M$, this implies that

$$a_N > a_M + r$$

for some positive real r, and so as $\{a_n\}$ is monotonic, the sentence

$$\exists n(n > m \wedge a_n > a_m + r)$$

is true in HR and hence in R for each finite positive integer m.

Thus, for example, there is an n_1 (finite) such that

$$a_{n_1} > a_1 + r,$$

an n_2 (finite) such that

$$a_{n_2} > a_{n_1} + r > a_1 + 2r,$$

an n_3 (finite) such that

$$a_{n_3} > a_{n_2} + r > a_1 + 3r,$$

and, proceeding in this way, it becomes clear that $\forall k$, if k is a positive integer, there exists a positive integer n such that

$$a_n > a_1 + kr$$

is true in R. As this sentence is also true in HR, we must have that

$$a_S > a_1 + Tr$$

for infinite integers S and $T > 0$. Since r is a positive real, Tr is infinite and so a_S must be infinite. This contradicts the assumption that $\{a_n\}$ is bounded, and our theorem follows. $\square$

With a mind toward eventually considering infinite polynomials, let us now examine the notion of infinite sums of numbers.

We do not listen with best regard to the verses of a man who is only a poet, nor to his problems if he is only an algebraist; but if a man is at once acquainted with the geometric foundation of things and with their festal splendor, his poetry is exact and his arithmetic musical.
Ralph Waldo Emerson (1803–1882)

The sum of infinitely many numbers is a more familiar idea than you probably think. Consider, for example, the real number

$.1\,1\,1\,1\,1\,1\,1\,\overline{1}\,\ldots.$

This is exactly the infinite sum

$$\frac{1}{10} + \frac{1}{100} + \frac{1}{1,000} + \frac{1}{10,000} + \cdots$$

You may recognize this as the fraction 1/9—we will prove this later.

In another situation, the Greek philosopher Zeno raised the following puzzling point: Suppose Achilles is chasing a tortoise. For simplicity, let us say that Achilles runs twice as fast as the tortoise, and that he starts one mile from it. Zeno then argues that Achilles will never catch the tortoise, for he must run to where the tortoise started—but by that time the tortoise is 1/2 mile beyond.

◄—— I MILE ——►

| | |
| I | 1/2 |

Achilles must then run to that point but again the tortoise has moved.

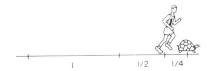

| | | |
| I | 1/2 | 1/4 |

We certainly know how to find, given any *two* numbers a_1 and a_2, their sum

$a_1 + a_2,$

or, in general, how to find, given any finitely many numbers $a_1, a_2, \ldots, a_n$, their sum

$a_1 + a_2 + \cdots + a_n.$

But what if we are given infinitely many numbers $a_1, a_2, \ldots, a_i, \ldots$. What do we then mean by their sum

$a_1 + a_2 + \cdots + a_i + \cdots?$

DEFINITION. A sum of infinitely many real numbers

$a_1 + a_2 + \cdots + a_i + \cdots$

is known as an *infinite series*. Let $\{S_n\}$ be the sequence of numbers where for each n

$$S_n = a_1 + a_2 + \cdots + a_n = \sum_{i=1}^{n} a_i.$$

Then if the sequence $\{S_n\}$ converges to a real number A, we say that that number is the sum

$a_1 + a_2 + \cdots + a_i + \cdots.$

In this case the infinite series

$a_1 + a_2 + \cdots + a_i + \cdots$

is said to *converge*. If S_N is infinite for some infinite N, or if there are infinite N_1 and N_2 such that S_{N_1} and S_{N_2} are not infinitely close to one another, then the infinite series

$a_1 + a_2 + \cdots + a_i + \cdots$

is said to *diverge*.

Notation: We will abbreviate the sum of the infinite series

$a_1 + a_2 + \cdots + a_i + \cdots$

by

$$\sum_{i=1}^{\infty} a_i.$$

Whenever

$a_1 + a_2 + \cdots + a_i + \cdots$

converges to a, we will write

In this way Achilles will never catch up with the tortoise because the tortoise has always moved beyond!

Analyzing this, suppose it takes Achilles t minutes to run one mile. Then it takes t minutes to run to where the tortoise started, $\frac{1}{2}t$ to run to the next point, $\frac{1}{4}t$ to run to the next, and so on. The mathematician can then say to Zeno that Achilles will catch the tortoise because the time it takes for Achilles to run these distances,

$$t + \frac{1}{2}t + \frac{1}{4}t + \frac{1}{8}t + \cdots,$$

is finite. That it *is* finite, we will prove shortly.

The most common and easily studied series is the geometric series. A geometric series is one of the form

$$1 + a + a^2 + a^3 + a^4 + \cdots$$

where a is some real number.

Here is a useful trick for finding the sum of a geometric series: Since

$$S_n = 1 + a + a^2 + \cdots + a^n,$$

$$aS_n = a + a^2 + \cdots + a^{n+1}.$$

By comparing the two right hand sides, we see that

$$S_n = aS_n + 1 - a^{n+1}.$$

Thus

$$S_n = \frac{1 - a^{n+1}}{1 - a}.$$

Now suppose $|a| < 1$. Then for any infinite $N > 0$, a^{N+1} is infinitesimal, and so

$$S_N = \frac{1 - a^{N+1}}{1 - a} \approx \frac{1}{1 - a}.$$

We have just proved that if $|a| < 1$, then the sum of $1 + a + a^2 + \cdots$ is $1/(1 - a)$.

In the case where $a = 1/10$, this tells us that

$$\sum_{i=1}^{\infty} a_i = a.$$

Essentially, our definition is

$$\sum_{i=1}^{\infty} a_i = a \qquad \text{iff} \qquad \sum_{i=1}^{N} a_i \approx a$$

for all infinite $N > 0$.

The following results cover some of the basic and useful facts about infinite series.

THEOREM 9.6. If $\sum_{i=1}^{\infty} a_i$ converges, then the infinite sequence

$$a_1, a_2, a_3, a_4, a_5, \cdots, a_i, \ldots$$

converges to 0.

PROOF: Let $S_k = a_1 + a_2 + \cdots + a_k$. Since the sequence S converges, it is Cauchy, and so $S_{K+1} - S_K \approx 0$ for each infinite K.

But $S_{K+1} - S_K = a_K$. Thus the sequence

$$a_1, a_2, \ldots, a_i, \ldots$$

converges to 0. $\square$

THEOREM 9.7. If $\sum_{i=1}^{\infty} a_i = a$ and $\sum_{i=1}^{\infty} b_i = b$, then $\sum_{i=1}^{\infty} a_i + b_i = a + b$ and $\sum_{i=1}^{\infty} a_i - b_i = a - b$. Furthermore, $\sum_{i=1}^{\infty} ka_i = ka$ for any real k.

PROOF: Exercise.

THEOREM 9.8. If $\sum_{i=1}^{\infty} a_i$ is a sum of nonnegative numbers, then it converges iff the sequence S given by

$$S_n = \sum_{i=1}^{n} a_i$$

is bounded.

PROOF: As each a_i is nonnegative, the sequence $\{S_n\}$ is monotonic. Thus by theorem 9.5 $\{S_n\}$ converges iff $\{S_n\}$ is bounded. $\square$

A useful test for convergence is the following:

THEOREM 9.9 (Comparison Test). If $a_1 + a_2 + \cdots + a_i + \cdots$ is a series such that each a_i is nonnegative, and if $b_1 + b_2 + \cdots + b_i + \cdots$ is a convergent series such that

$$a_i \leq b_i \qquad \text{for all } i,$$

then $a_1 + a_2 + \cdots + a_i + \cdots$ converges.

$$1 + \frac{1}{10} + \frac{1}{100} + \frac{1}{1,000} + \cdots = \frac{1}{1 - 1/10}$$

$$= \frac{10}{9},$$

and so

$$.1\,1\,1\,1\,1\,1\,\overline{1} = \frac{1}{10} + \frac{1}{100} + \frac{1}{1,000} + \cdots$$

$$= \frac{10}{9} - 1$$

$$= \frac{1}{9}.$$

Is the converse of this theorem true? If a series converges, does it converge absolutely? The answer is no. The series

$$1 - \frac{1}{2} + \frac{1}{3} - \frac{1}{4} + \frac{1}{5} - \cdots$$

converges, but not absolutely. We shall see this shortly in this column.

In high school you may have been taught the following method: Let $x = .11\overline{1}\ldots$. Then $10x = 1.11\overline{1}\ldots$, and so $10x - x = 9x = (1.11\overline{1}\ldots) - (.111\overline{1}\ldots) = 1$. Thus $x = 1/9$. In doing this, we are using theorem 9.7, since we are subtracting two infinite series from each other.

Using theorem 9.7 we can resolve Zeno's paradox rigorously. By the theorem,

$$t + \frac{1}{2}t + \frac{1}{4}t + \frac{1}{8}t + \cdots$$

$$= t\left(1 + \frac{1}{2} + \frac{1}{4} + \frac{1}{8} + \cdots\right)$$

$$= t\left(1 + \frac{1}{2} + \left(\frac{1}{2}\right)^2 + \left(\frac{1}{2}\right)^3 + \cdots\right)$$

$$= t\left(\frac{1}{1 - 1/2}\right).$$

Thus the infinite sum

$$t + \frac{1}{2}t + \frac{1}{4}t + \frac{1}{8}t + \cdots$$

PROOF: Let $S_n = \sum_{i=1}^{n} b_i$ and $T_n = \sum_{i=1}^{n} a_i$. By theorem 9.8, the sequence S_n is bounded. By hypothesis, $0 \leq T_n \leq S_n$, hence T_n is bounded. Thus (again by theorem 9.8) $\sum_{i=1}^{\infty} a_i$ converges. $\square$

Note: The contrapositive of the comparison test is a useful test for divergence: If for all i, $0 < a_i \leq b_i$, then if $\sum_{n=1}^{\infty} a_n$ diverges, $\sum_{n=1}^{\infty} b_n$ diverges.

A more powerful, though more restrictive, notion of convergence is absolute convergence.

DEFINITION. The series $a_1 + a_2 + \cdots + a_i + \cdots$ is said to be *absolutely convergent* iff the associated series

$$|a_1| + |a_2| + \cdots + |a_i| + \cdots$$

converges.

THEOREM 9.10. If $a_1 + a_2 + \cdots + a_i + \cdots$ is absolutely convergent, then it is convergent.

PROOF: We define two new series:

(1) Let $a_n^+ = \begin{cases} a_n & \text{if } a_n \geq 0 \\ 0 & \text{if } a_n < 0. \end{cases}$

(2) Let $a_n^- = \begin{cases} -a_n & \text{if } a_n \leq 0 \\ 0 & \text{if } a_n > 0. \end{cases}$

Since each a_n^+ and a_n^- is less than

$$|a_n|,$$

the comparison test tells us that

$$\sum_{n=1}^{\infty} a_n^+ \quad \text{and} \quad \sum_{n=1}^{\infty} a_n^-$$

converge. But for each n, $a_n^+ - a_n^- = a_n$, and so by theorem 9.6,

$$\sum_{n=1}^{\infty} a_n = \sum_{n=1}^{\infty} a_n^+ - \sum_{n=1}^{\infty} a_n^-$$

converges. $\square$

The following is perhaps the most useful test for convergence:

THEOREM 9.11 (Ratio Test). Let $a_1 + a_2 + \cdots + a_i + \cdots$ be a given series with $a_i \neq 0$ for all i. Assume that the ratio

$$\frac{|a_{N+1}|}{|a_N|}$$

is finite and has the same standard part r for all infinite

is finite and equals

$$t\left(\frac{1}{1-1/2}\right) = 2t.$$

Convert these decimals to fractions.

1. $.22\bar{2}...$
2. $.1212\overline{12}...$ *Hint*: $.12\overline{12}...$ is

$$\frac{12}{100} + \frac{12}{10,000} + \frac{12}{1,000,000} + \cdots$$

$$= \frac{12}{100}\left(1 + \frac{1}{100} + \left(\frac{1}{100}\right)^2 + \cdots\right).$$

3. $.3737\overline{37}...$
4. $.456456\overline{456}...$
5. $.31111\bar{1}...$
6. $.538242\overline{42}...$

The modern mathematics student should be quite familiar with numbers written in different bases. For example,

$$.243_6 = \frac{2}{6} + \frac{4}{36} + \frac{3}{216},$$

$$.011_2 = \frac{0}{2} + \frac{1}{4} + \frac{1}{8},$$

and

$$.6663_7 = \frac{6}{7} + \frac{6}{49} + \frac{6}{343} + \frac{3}{2,401}.$$

Convert the following numbers to fractions:

7. $.11\bar{1}..._2$
8. $.22\bar{2}..._7$
9. $.13\bar{13}..._4$
10. $.6422\bar{2}..._9$

It is also possible to have *negative* number bases. For example,

$$.243_{-6} = \frac{2}{-6} + \frac{4}{36} + \frac{3}{-216},$$

$$.1011_{-2} = \frac{1}{-2} + \frac{0}{4} + \frac{1}{-8} + \frac{1}{16}.$$

Convert the following numbers to fractions:

11. $.222\bar{2}..._{-5}$
12. $1.111\bar{1}..._{-2}$
13. $.53\bar{53}..._{-6}$
14. $0.22\bar{2}..._{-10}$

It is actually possible to express any real number, positive or negative, in a negative-number base.

$N \geq 0$. Then (1) if $r < 1$, the series converges absolutely, and (2) if $r > 1$, the series diverges.

PROOF: (1) We must show the series $\sum_{n=1}^{\infty} |a_n|$ convergent. Since

$$\left\|\frac{a_{N+1}}{a_N}\right\| = r < 1$$

for all infinite N, there is a real number t, $0 < r < t < 1$, such that

$$\exists n_0 \forall n\left(\text{if } n > n_0, \text{ then } \left|\frac{a_{n+1}}{a_n}\right| < t\right)$$

is true in HR (take n_0 infinite). Thus

$$\exists n_0 \forall n\left(\text{if } n > n_0, \text{ then } \left|\frac{a_{n+1}}{a_n}\right| < t\right)$$

is true in **R**, and so

$$|a_{n+1}| < t\,|a_n|$$

for all finite n larger than some n_0. This immediately implies

$$|a_{n_0+2}| < t\,|a_{n_0+1}| < t^2\,|a_{n_0}|,$$

$$|a_{n_0+3}| < t\,|a_{n_0+2}| < t^3\,|a_{n_0}|,$$

and, in general,

$$|a_j| < t^{j-n_0}\,|a_{n_0}|$$

for all $j > n_0$. Thus

$$\sum_{i=1}^{\infty} |a_i| = \sum_{i=1}^{n_0} |a_i| + \sum_{j=n_0+1}^{\infty} |a_j|$$

$$< \sum_{i=1}^{n_0} |a_i| + \sum_{j=n_0+1}^{\infty} t^{j-n_0}\,|a_{n_0}|$$

$$< \sum_{i=1}^{n_0} |a_i| + |a_{n_0}| \sum_{j=1}^{\infty} t^j.$$

As $0 < t < 1$, $\sum_{j=1}^{\infty} t^j$ converges and hence $\sum_{i=1}^{\infty} |a_i|$ converges.

(2) Since $\left\|\dfrac{a_{N+1}}{a_N}\right\| > 1$ for all infinite N,

$$\exists n_0 \forall n\left(\text{if } n > n_0, \text{ then } \left|\frac{a_{n+1}}{a_n}\right| > 1\right)$$

is true in HR (take n_0 infinite). Thus

An interesting new way to write numbers has recently been suggested and given the name *fracimals*. Some examples are

$$.2.3. = \frac{1}{2} + \frac{1}{2\cdot3},$$

$$.7.11.5. = \frac{1}{7} + \frac{1}{7\cdot11} + \frac{1}{7\cdot11\cdot5},$$

$$.2.5.3.2. = \frac{1}{2} + \frac{1}{2\cdot5} + \frac{1}{2\cdot5\cdot3}$$
$$+ \frac{1}{2\cdot5\cdot3\cdot2}.$$

There are some curious duplications.

$$.2. = \frac{1}{2},$$

$$.3.2. = \frac{1}{3} + \frac{1}{3\cdot2} = \frac{1}{2},$$

$$.3.3.2. = \frac{1}{3} + \frac{1}{3\cdot3} + \frac{1}{3\cdot3\cdot2} = \frac{1}{2}.$$

This suggests that perhaps the infinite fracimal:

$$.3.3.3.3.3... = \frac{1}{2}.$$

EXERCISE

1. Prove that $.3.3.3.... = \frac{1}{2}$.

Write these fracimals as fractions:
2. .4.4.4.4...
3. .2.3.2.3.2.3...
4. .5.3.3.3.3.3...

What is the sum of the series

$$1 + \frac{1}{2} + \frac{1}{3} + \frac{1}{4} + \frac{1}{5} + \frac{1}{6} + \cdots ?$$

We shall show that it is not convergent. Consider the series

$$1 + \frac{1}{2} + \frac{1}{4} + \frac{1}{4} + \frac{1}{8} + \frac{1}{8} + \frac{1}{8} + \frac{1}{8}$$

$$+ \frac{1}{16} + \frac{1}{16} + \frac{1}{16} + \frac{1}{16} + \frac{1}{16}$$

$$+ \frac{1}{16} + \frac{1}{16} + \frac{1}{16} + \frac{1}{32} + \cdots.$$

This series is unbounded, for if S_n is the sum of the first n terms,

$$\exists n_0 \forall n \left(\text{if } n > n_0 \text{ then } \left| \frac{a_{n+1}}{a_n} \right| > 1 \right)$$

is true in **R**, and so

$$|a_{n+1}| > |a_n|$$

for finite n larger than some n_0. Thus for N infinite,

$$|a_N| > |a_{n_0}| > 0,$$

and so $a_N \not\approx 0$, and thus by theorem 9.8,

$$\sum_{n=1}^{\infty} a_n$$

diverges. ☐

Consider the series

$$\sum_{n=1}^{\infty} \frac{1}{n^2}.$$

Does it converge? If we apply the ratio test,

$$\frac{a_{N+1}}{a_N} = \frac{(N+1)^2}{N^2}$$

$$= \frac{N^2 + 2N + 1}{N^2}$$

$$= 1 + \frac{2}{N} + \frac{1}{N^2}$$

$$= 1,$$

we learn nothing. For this series we need an additional test.

THEOREM 9.12 (Integral Test). Suppose a positive sequence $\{a_n\}$ is given by a function f,

$$f(n) = a_n,$$

and suppose this f is monotonic decreasing. Then if $\int_1^N f(x)dx$ is finite, $\sum_{n=1}^{\infty} a_n$ converges.

we have $S_1 = 1$, $S_2 = 1\frac{1}{2}$, $S_4 = 2$, $S_8 = 2\frac{1}{2}$, $S_{16} = 3$, and so on. In general,

$$S_{2^n} = 1 + \frac{n}{2}.$$

For infinite $N > 0$,

$$S_{2^N} = 1 + \frac{N}{2},$$

which is not finite. By theorem 9.8, therefore, this series diverges. By the contrapositive to theorem 9.9 (p. 90), this implies that

$$1 + \frac{1}{2} + \frac{1}{3} + \frac{1}{4} + \cdots \text{ diverges.}$$

A consequence of this is that the sequence

$$1 - \frac{1}{2} + \frac{1}{3} - \frac{1}{4} + \frac{1}{5} - \cdots$$

does not converge absolutely. We will see in the next chapter that it does converge.

EXERCISES

Discover whether these series converge or diverge:

1. $\sum_{i=1}^{\infty} \frac{1}{i!}$ 2. $\sum_{i=1}^{\infty} \frac{2}{i!}$ 3. $\sum_{i=1}^{\infty} \frac{2}{n}$

4. $\sum_{i=1}^{\infty} \frac{1}{2n}$ 5. $\sum_{i=1}^{n} \sin \pi n$

6. $\sum_{i=1}^{\infty} \cos \pi n$ 7. $\sum_{i=1}^{\infty} \frac{i!}{1,000,000^i}$

8. $\sum_{i=1}^{\infty} \frac{i}{2^i}$ 9. $\sum_{i=1}^{\infty} \frac{i^2}{2^i}$

10. $\sum_{i=1}^{\infty} \frac{1}{\sqrt{i!}}$ 11. $\sum_{i=1}^{\infty} \frac{1}{i^i}$

12. $\sum_{i=1}^{\infty} \frac{x^i}{i!}$, where x is any real number.

PROOF: Graphically, the sum of the series $\sum_{n=1}^{N} a_n$ is the area of the boxes:

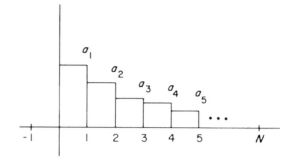

On the other hand, the integral $\int_1^N f(x)dx$ is:

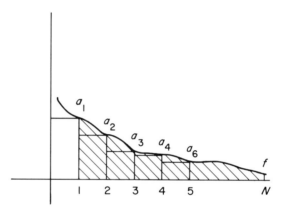

Thus, if $\int_1^N f(x)\,dx$ is finite, $\sum_{n=1}^{N} a_n$ is finite, and so it converges by theorem 9.8. $\square$

Applying this to $\{1/n^2\}$,

$$\int_1^N \frac{dx}{x^2} = \frac{-1}{x}\Big|_1^N = \frac{-1}{N} + 1,$$

which is finite.

EXERCISES

Which of these series converge?

1. $\sum_{i=1}^{\infty} \frac{1}{i^3}$.   2. $\sum_{i=1}^{\infty} \frac{1}{i\sqrt{i}}$.

3. Prove the converse of the theorem, that if $\int_1^N f(x)dx$ is infinite, then $\sum_{n=1}^{N} a_n$ is infinite. (*Hint:* Move the boxes over one unit so the curve lies inside them.)

4. Use the theorem and its converse to discover for which real numbers p, $\sum_{i=1}^{\infty} 1/n^p$ converges.

It is true that the mathematician who is not somewhat of a poet, will never be a perfect mathematician.
Karl Theodor Weierstrass
(1815–1897)

Now that we have developed machinery for dealing with infinite series we are ready to readdress our original problem, that of trying to view arbitrary functions as infinitely long *polynomials*.

For a start, since every polynomial is successively differentiable, our only candidates for functions which might be equal to infinite polynomials are functions that are differentiable, whose derivatives are differentiable, and so forth ad inf. (Such functions will henceforth be known as "infinitely differentiable.")

Suppose that we are given such an infinitely differentiable function, say $f(x)$.

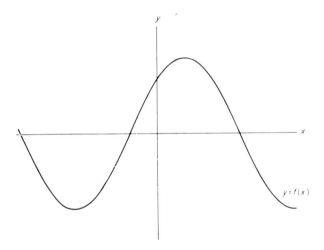

How would we go about trying to find a polynomial that was equal to $f(x)$? We can try approximating it by stages: Let $p_0(x)$ be the polynomial

$$y = f(0):$$

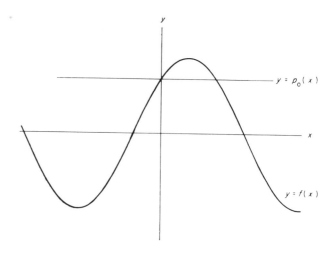

$p_0(x)$ is a polynomial that equals $f(x)$ for $x = 0$ but doesn't seem to approximate $f(x)$ too well for $x \neq 0$. Next let $p_1(x)$ be the polynomial

$$y = f(0) + f'(0)x:$$

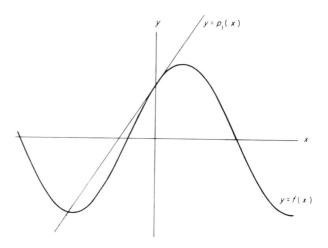

$p_1(x)$ is a polynomial that equals $f(x)$ at $x = 0$ and whose first derivative equals that of $f(x)$ at $x = 0$. As such, it is fair to say that $p_1(x)$ approximates $f(x)$ better than $p_0(x)$, for not only does $p_1(x)$ equal $f(x)$ at $x = 0$, but the direction in which the graph of p_1 leaves $(0, f(0))$ is the same direction in which the graph of $f(x)$ leaves $(0, f(0))$, that direction being $f'(0)$.

We can carry this approximation process even further: Let $p_2(x)$ be the polynomial

$$f(0) + f'(0)x + \tfrac{1}{2} f''(0)x^2.$$

Then $p_2(x)$ is a polynomial that equals $f(x)$ at $x = 0$ and whose first and second derivatives equal those of $f(x)$ at $x = 0$. As such, it can be said that $p_2(x)$ approximates $f(x)$ even better than $p_0(x)$ or $p_1(x)$,

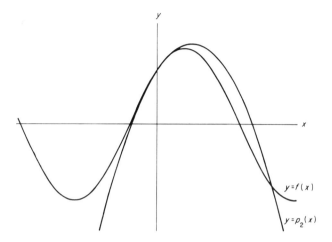

for since the second derivative of $p_2(x)$ at $x = 0$ equals that of $f(x)$, the slope of $p_2(x)$ at $x = 0$ is changing in the same way the slope of $f(x)$ is, that is, $p_2(x)$ is curving the same way $f(x)$ is at $x = 0$.

We can continue approximating in this way ad infinitum. $p_3(x)$ would be

$$f(0) + f'(0)x + \frac{f''(0)}{2} x^2 + \frac{f'''(0)}{2 \cdot 3} x^3,$$

$p_4(x)$ would be

$$f(0) + f'(0) x + \frac{f''(0)}{2} x^2 + \frac{f'''(0)}{2 \cdot 3} x^3 + \frac{f''''(0)}{2 \cdot 3 \cdot 4} x^4,$$

and in general, $p_n(x)$ would be

$$\sum_{k=0}^{n} \frac{f^{(k)}(0)}{k!} x^k.$$

The question naturally arises: Is $p_n(x)$ actually *equal* to $f(x)$ to some n? Of course, the answer depends upon just which function $f(x)$ is, but for most functions $f(x)$ the answer is NO!

However, we have only looked at *finite* polynomial approximations to $f(x)$, that is, at polynomials of the form

$$\sum_{k=0}^{n} \frac{f^{(k)}(0)}{k!} x^k$$

where n is a finite natural number. What about poly-

An example:

$$f(x) = \sqrt{x + 1}\cdot$$

Computing derivatives:

$$f'(x) = \frac{1}{2\sqrt{x + 1}},$$

$$f''(x) = -\tfrac{1}{4}(x + 1)^{-3/2},$$

$$f'''(x) = \tfrac{3}{8}(x + 1)^{-5/2}.$$

Thus $f(0) = 1,$

$$f'(0) = \tfrac{1}{2},$$

$$f''(0) = -\tfrac{1}{4},$$

$$f'''(0) = \tfrac{3}{8},$$

and the polynomial $p_\infty(x)$ is

$$1 + \frac{x}{2} - \frac{x^2}{8} + \frac{x^3}{16}\cdots$$

By inspection, the nth term is

$$\pm \frac{x^{n-1}}{2^n}.$$

By the ratio test, $p_\infty(x)$ is always convergent.

Another example:

$$f(x) = \exp x = e^x.$$

Since

$$f(x) = f'(x) = f''(x) = \ldots,$$

we have

$$1 = f(0) = f'(0) = f''(0) = \ldots,$$

and $p_\infty(x)$ is

$$1 + x + \frac{x^2}{2} + \frac{x^3}{6} + \frac{x^4}{24} + \cdots.$$

Taylor's Theorem says that the difference between $f(x)$ and $p_n(x)$ is

$$\frac{e^c x^{n+1}}{(n + 1)!}.$$

We now show that this is an infinitesimal when N is infinite. Since

$$\sum_{i=1}^{\infty} \frac{x^i}{i!}$$

converges by the ratio test (exer. 12, p. 94),

nomials of the form

$$p_\infty(x) = \sum_{k=0}^{\infty} \frac{f^{(k)}(0)}{k!} x^k?$$

Such "infinite polynomials" are usually known as *power series* and are functions defined as follows: For any real number r, $p_\infty(r)$ is the sum of the infinite series

$$\sum_{k=0}^{\infty} \frac{f^{(k)}(0)r^k}{k!}$$

provided the series converges; otherwise $p_\infty(r)$ is undefined. Note that $p_\infty(r)$ will equal a real number b exactly when $p_N(r) \approx b$ for every infinite $N > 0$. Thus, to attack the question of whether $f(x) = p_\infty(x)$, we may use the following well-known theorem of Taylor:

THEOREM 10.1. Let f be a given infinitely differentiable function and let n be a nonnegative integer. Then for any b the difference between $f(b)$ and the polynomial $p_n(b)$,

$$f(b) - p_n(b),$$

is equal to

$$\frac{f^{(n+1)}(c)}{(n + 1)!} b^{n+1}$$

where c is some number lying between b and 0. (*Note:* This theorem applies whether n is finite or infinite.)

Before actually proving this theorem, let us see how it helps us to answer our original question with an example.

Consider the function $f(x) = \sin x$. We know that

$$f'(x) = \cos x,$$

$$f''(x) = -\sin x, \text{ etc.}$$

Thus, since $\sin 0 = 0$ and $\cos 0 = 1$,

$$p_1(x) = x,$$

$$p_2(x) = x,$$

$$p_3(x) = x - \frac{x^3}{3!},$$

$$p_4(x) = x - \frac{x^3}{3!},$$

$$p_5(x) = x - \frac{x^3}{3!} + \frac{x^5}{5!}, \text{ etc.}$$

$$\frac{x^{N+1}}{(N + 1)!}$$

is infinitesimal by theorem 9.6.
Thus for N infinite,
$x^{N+1}/(N + 1)!$ is infinitesimal, and so
$e^c x^{N+1}/(N + 1)!$ is infinitesimal (e^c is
finite, lying between e^0 and e^x).
We can now conclude that

$$f(x) \approx f_N(x),$$

or

$$e^x = \sum_{i=1}^{\infty} \frac{x^{i-1}}{(i - 1)!}$$

(0! is defined to be 1).

This formula is very valuable for
approximating e^x. On page 83 we
needed, for example, a value for $e^{.06}$
correct to seven significant digits. We
found this using the above series. We
let $x = .06$ and carried out the opera-
tion as far as necessary:

$$e^{.06} = 1 + .06 + \frac{(.06)^2}{2} + \frac{(.06)^3}{6} + \cdots$$

$$= 1$$
$$+ .06$$
$$+ .0018$$
$$+ .000036$$
$$+ .00000054$$
$$\vdots$$

and it is clear that 1.061837 is $e^{.06}$ to
that accuracy. (*Note*: These operations were carried out on a pocket
calculator.)

EXERCISES
Find the first five terms of the
MacLaurin's series for the following
functions:

1. $f(x) = \cos x$.

2. $f(x) = \dfrac{1}{1 - x}$.

3. $f(x) = \dfrac{1}{1 + x}$.

4. $f(x) = e^{2x}$.

5. $f(x) = \sin 2x$.

6. $f(x) = \log(1 + x)$.

Do we ever have $f(x) = p_n(x)$? By Taylor's Theorem,
for any x,

$$f(x) - p_n(x) = \frac{f^{(n+1)}(c)}{(n + 1)!} x^{n+1}$$

for some c between x and 0, and as, in this case,

$$-1 \leq f^{(n+1)}(c) \leq 1$$

for *every* c, if $N > 0$ is an infinite integer,

$$\frac{f^{(N+1)}(c)}{(N + 1)!} x^{N+1}$$

is infinitesimal for every x lying between 1 and -1. (It
is actually infinitesimal for *every* x but this is not obvious—see left column.)

We can thus immediately say that for $-1 \leq x \leq 1$,
the difference between $f(x) = \sin x$ and

$$p_N(x) = x - \frac{x^3}{3!} + \frac{x^5}{5!} - \cdots + \frac{x^{2N+1}}{(2N + 1)!}$$

is infinitesimal. Thus $f(x) = p_\infty(x)$.
The series

$$f(0) + f'(0)x + \frac{f''(0)x^2}{2} + \cdots$$

is called the *MacLaurin's series* for f.

As we shall see shortly, this ability to think of functions as polynomials, even infinite polynomials, is
extremely useful.

PROOF: The proof of Taylor's Theorem is quite easy
and has essentially been given already in an earlier
chapter. Indeed, recall that the Mean Value Theorem
tells us that for some c between b and 0,

$$f'(c) = \frac{f(b) - f(0)}{b - 0}$$

or, in our current notation, that

$$f(b) - f(0) = f'(c)b.$$

This is Taylor's Theorem for $n = 0$.

We proved the Mean Value Theorem as an immediate consequence of Rolle's Theorem: If $f(a) = f(b) =$
0, then for some c, $a < c < b$, $f'(c) = 0$; and similarly
we can immediately derive Taylor's Theorem from a

7. $f(x) = \dfrac{\sin x}{x}$.

8. $f(x) = \dfrac{1}{1 - x^2}$.

9. Having done exercises 4–8 the hard way, see if you can spot the easy way by using previous examples and exercises.

Taylor's Theorem can be used to prove an interesting fact: e is irrational. From the series, we have that

$$e = e^1 = 1 + 1 + \frac{(1)^2}{2} + \frac{(1)^3}{6} + \cdots$$

$$= 1 + 1 + 1/2 + 1/6 + 1/24 + \cdots$$

$$= \sum_{i=0}^{\infty} \frac{1}{i!},$$

and Taylor's Theorem states that the difference

$$e - \sum_{i=0}^{N} \frac{1}{i!} = \frac{f^{N+1}(c)(1)^{N+1}}{(N+1)!}$$

or

$$e - \left(1 + 1 + \cdots + \frac{1}{N!}\right) = \frac{e^c}{(N+1)!},$$

where c is between 0 and 1. Now we claim that this proves e is irrational, for if e were rational, then $N!e$ must be an integer. But then if we multiply the equation above by $N!$, we get an integer on the left-hand side, and

$$\frac{e^c}{N+1},$$

an infinitesimal, on the right. Since this is impossible, e cannot be rational.

EXERCISES
Use previous exercises and examples to approximate the following numbers to the nearest hundredth:
1. $e^{.13}$.
2. $\sin .32$.
3. $\sqrt{1.07}$.
4. $\cos \sqrt{3}$.
5. $\log .97$.
6. Find the series for $(1 + x)^{2/5}$ and calculate $(1.3)^{2/5}$ to the nearest hundredth.
7. Prove that the fracimal .1.2.3.4.5.6 ... = $e - 1$ (see p. 93).

generalized Rolle's Theorem: If $f(a) = f(b) = 0$ and $f'(a) = f''(a) = \cdots = f^{(n)}(a) = 0$, then for some c, $a < c < b, f^{(n+1)}(c) = 0$. But this generalized Rolle's Theorem follows immediately from the original Rolle's Theorem: As $f(a) = f(b) = 0, f'(c_1) = 0$ for some $a < c_1 < b$, hence

as $f'(a) = f'(c_1) = 0, f''(c_2) = 0$ for some $a < c_2 < c_1$, hence

as $f''(a) = f''(c_2) = 0, f'''(c_3) = 0$ for some $a < c_3 < c_2$, hence

etc., hence $f^{(n+1)}(c) = 0$ for some $a < c < c_n < b$.

The proof now of Taylor's Theorem from this generaized Rolle's theorem is entirely similar to the derivation of the Mean Value Theorem from the original Rolle's Theorem: we simply "rotate" the picture.

More specifically, let our "rotated" version of $f(x)$, $g(x)$, be the function defined as follows:

$$(*) \quad g(x) = f(x) - p_n(x) + \left(\frac{x}{b}\right)^{n+1} (p_n(b) - f(b)).$$

The reader can routinely verify that

$$g(0) = g(b) = 0 \tag{1}$$

and

$$g'(0) = g''(0) = \cdots = g^n(0) = 0. \tag{2}$$

Thus, by our generalized Rolle's Theorem, there is a number c between 0 and b such that

$$(**) \quad g^{(n+1)}(c) = 0.$$

8. Prove that $.2.4.6.8.10.12... = \sqrt{e} - 1$.

9. Those readers who have been introduced to the special number i, whose important property is that $i^2 = -1$, can use series to prove Demoivre's Theorem:

$$e^{ix} = \cos x + i\sin x.$$

Taylor's Theorem solves another problem for us. Consider the series for the function

$$f(x) = \ln(1 + x).$$

If you successfully did exercise 6 on p. 99, you found the series

$$x - \frac{x^2}{2} + \frac{x^3}{3} - \frac{x^4}{4} + \frac{x^5}{5} - \cdots.$$

We are interested in this series for $x = 1$:

$$1 - \tfrac{1}{2} + \tfrac{1}{3} - \tfrac{1}{4} + \tfrac{1}{5} - \cdots,$$

for we promised in Chapter 9 to prove this series convergent (see p. 94). By Taylor's Theorem, in order to show that

$$\ln(1 + 1) = 1 - \tfrac{1}{2} + \tfrac{1}{3} - \tfrac{1}{4} + \tfrac{1}{5} - \cdots,$$

we need only show that

$$\left| \frac{f^{N+1}(c)1^{N+1}}{(N + 1)!} \right|$$

is infinitesimal for all c between 0 and 1. However,

$$|f^{N+1}(c)| = N!(1 + c)^{-(N+1)}$$

and so

$$\left| \frac{f^{N+1}(c)1^{N+1}}{(N + 1)!} \right| = \frac{1}{(N+1)(1+c)^{N+1}},$$

an infinitesimal.

If we calculate $g^{(n+1)}(c)$ using (*), the equation (**) becomes

$$f^{n+1}(c) + \frac{(n + 1)!}{b^{n+1}}(p_n(b) - f(b)) = 0$$

or

$$f(b) - p_n(b) = \frac{f^{n+1}(c)b^{n+1}}{(n + 1)!}. \quad \square$$

Remark: Our just concluded work concerned itself exclusively with polynomial approximations centered at $x = 0$:

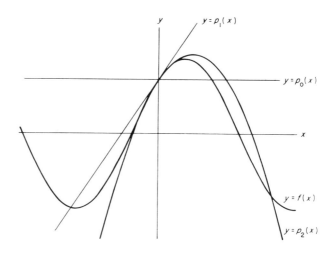

There are times, however, when it is more convenient to center ourselves at $x = a \neq 0$:

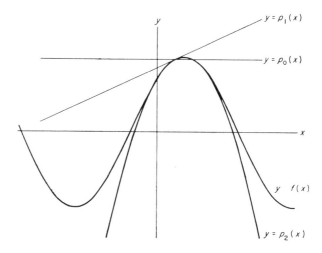

Consider the following plan for increasing the male population of the world. All families continue to have children as long as they are boys. As soon as the first girl is born they must have no more children. Will this plan succeed? Of course not! Regardless of any plan, roughly half of all births will be male, half female. To see this another way, let us calculate the number of children born under this system. Suppose there are exactly P families. Under the law, half of these families will have just one child because their first will be a girl. That's $1 \cdot P/2$ children.

Of those that are left, half will have to stop after two children, for their second will be a girl. That's $P/4$ families with two children each, or $2 \cdot P/4$ more children.

Of those that are left, half will have to stop after 3 children—$P/8$ families, 3 kids each, or $3 \cdot P/8$.

Continuing in this way we get an infinite series:

$$1 \cdot P/2 + 2 \cdot P/4 + 3 \cdot P/8 + 4 \cdot P/16 + \cdots.$$

Taking out $P/2$, we have

$$P/2(1 + 2(\tfrac{1}{2}) + 3 \cdot (\tfrac{1}{2})^2 + 4 \cdot (\tfrac{1}{2})^3 + \cdots).$$

This series is unlike any we have studied. When $\tfrac{1}{2}$ is replaced by x, it is a power series:

$$P/2(1 + 2x + 3x^2 + 4x^3 + 5x^4 + \cdots).$$

Even though it is new to us, we should recognize it as the derivative, term by term, of the geometric series

$$1 + x + x^2 + x^3 + x^4 + \cdots$$

which we know to equal

$$\frac{1}{1-x}.$$

Does this mean that

$$1 + 2x + 3x^2 + 4x^3 + \cdots$$

is the derivative of

$$\frac{1}{1-x},$$

that is, $1/(1-x)^2$? The answer is yes, and that is precisely theorem 10.4.

In this case all of our work carries over without a hitch. In the more general context, given any infinitely differentiable function h, we have that for each n,

$$h_n(x) = h(a) + h'(a)(x-a) + h''(a)\frac{(x-a)}{2!} + \cdots$$
$$+ h^{(n)}(a)\frac{(x-a)^n}{n!}$$

and

$$h(b) - h_n(b) = h^{(n+1)}(c)\frac{(b-a)^{n+1}}{(n+1)!}$$

for some c between b and a.

The series

$$\sum_{k=0}^{\infty} \frac{h^{(k)}(a)(x-a)^k}{k!},$$

where a is any real number, is called the *Taylor's series*. It is identical to the MacLaurin's series when $a = 0$.

We have thus far seen how functions which at first glance seem quite different from polynomials can be expressed as "*infinite* polynomials." This turns out to be quite useful, for as we shall shortly see, it is almost as easy to work with infinite polynomials (for example, adding them, multiplying them, differentiating them, integrating them) as it is to work with ordinary finite polynomials. But even more important is this: The functions that we are used to working with in elementary mathematics—polynomials, trigonometric functions, exponential and logarithmic functions, and algebraic combinations of these functions—make up an extremely small subcollection of the real-valued functions. In fact by using various more advanced techniques from mathematics one can show that "most" real-valued functions cannot be expressed in terms of elementary functions, and these nonelementary functions include many that arise naturally in physics, engineering, or the social sciences. This is where infinite polynomials have their greatest use.

We will now prove a number of theorems that demonstrate that infinite polynomials behave like finite ones.

THEOREM 10.2.
(1) If $p(r) = \sum_{i=0}^{\infty} a_i x^i$ converges at r, then it converges at any x such that $|x| < r$, and in fact it converges absolutely for such x.

Thus

$$1 + 2(\tfrac{1}{2}) + 3(\tfrac{1}{2})^2 + \cdots$$

$$= \frac{1}{(1 - \tfrac{1}{2})^2}$$

$$= 4.$$

And so the total number of children is

$$(P/2) \cdot 4 = 2P.$$

Now, since each family has exactly *one* girl, there are P girls, and hence, P boys.

Theorems 10.4 and 10.5 show that we can integrate and differentiate power series term-by-term. This is a tremendous help for certain functions that cannot be integrated any other way. Take for example $\sin x / x$. There is no "simple" function that is a primitive of this function. If we wish to find, say,

$$\int_0^2 \frac{\sin x}{x}\, dx$$

we can find it to any accuracy using series

$$\int_0^2 \frac{\sin x}{x}\, dx$$

$$= \int_0^2 \left(1 - \frac{x^2}{3!} + \frac{x^4}{5!} - \frac{x^6}{7!} + \cdots\right) dx$$

(exer. 7, p. 100)

$$= x - \frac{x^3}{18} + \frac{x^5}{600} - \frac{x^7}{35180} + \cdots \Big|_0^2$$

$$= 2 - .44444\ldots + .05333\ldots$$
$$\quad - .00363\ldots + \cdots$$

$$= 1.61 \text{ to the nearest hundredth.}$$

EXERCISE
Find the integral

$$\int_0^{1/2} e^{-x^2} dx$$

to the nearest thousandth.

Our method for proving theorem 10.4 is due to T. Apostol.

(2) If $\sum_{i=0}^{\infty} a_i x^i$ diverges for $x = r$, then it diverges for all x such that $|x| > |r|$.

PROOF OF (1): Since $\sum_{i=0}^{\infty} a_i r^i$ converges, $a_i r^i$ approaches 0 (theorem 9.8). Thus, it is bounded, say by M. We then have that for each i,

$$\left| a_i x^i \right| = \left| a_i r^i \right| \cdot \left| \frac{x^i}{r^i} \right| \le M \left| \frac{x}{r} \right|^i.$$

Since $\sum_{i=0}^{\infty} M |x/r|^i$ converges (by the ratio test), $\sum_{i=0}^{\infty} |a_i x^i|$ converges by the comparison test.

PROOF OF (2). If $\sum_{i=0}^{\infty} a_i x^i$ converged, then $\sum_{i=0}^{\infty} a_i r^i$ would also converge (by (1) above). This is a contradiction. $\square$

Let us now look at some calculus:

THEOREM 10.3. If $P(x) = \sum_{i=0}^{\infty} a_i x^i$ converges in $(-r, r)$, then so does $Q(x) = \sum_{i=0}^{\infty} i a_i x^{i-1}$.

PROOF: For any c, $0 \le |c| < r$, let b be such that

$$0 \le |c| < b < r.$$

Applying the Mean Value Theorem to the functions $a_i x^i$, for all i, we obtain real numbers d_i, $|c| < d_i < b$, such that

$$i a_i d_i^{i-1} = \frac{a_i |c|^i - a_i b^i}{|c| - b}.$$

Thus

$$\sum_{i=0}^{\infty} \left| i a_i d_i^{i-1} \right| = \frac{1}{|c| - b} \sum_{i=0}^{\infty} \left| (a_i |c|^i - a_i b^i) \right|.$$

Since $\sum_{i=0}^{\infty} a_i |c|^i$ and $\sum_{i=0}^{\infty} a_i b^i$ converge by hypothesis, so does $\sum_{i=0}^{\infty} i a_i d_i^{i-1}$; hence $\sum_{i=0}^{\infty} |i a_i c^i|$ converges by the comparison test. $\square$

THEOREM 10.4. If $f(x) = \sum_{i=0}^{\infty} a_i x^i$ converges in $(-r, r)$, then $f'(x) = \sum_{i=0}^{\infty} i a_i x^{i-1}$ in $(-r, r)$.

PROOF: Let $F(x) = \sum_{i=0}^{\infty} i a_i x^i$. For any c in $(-r, r)$, let $b \ne c$ also be in $(-r, r)$. Choose s such that $r > |s| > |b|, |c|$. As before, let d_i be between b and c, satisfying

$$\frac{a_i c^i - a_i b^i}{c - b} = i a_i d_i^{i-1}.$$

Apply the Mean Value Theorem once more to the functions x^{i-1} and obtain reals e_i between c and d_i

Perhaps the most sensational application of series is the solution of differential equations. In a differential equation information is given about an unknown function and its derivatives. Solving the equation consists of finding all functions that satisfy the equation. For example,

$$f'(x) = x.$$

Solving the equation is a simple case of integration. The functions which satisfy the equation are

$$x^2/2, \; x^2/2 + 1, \; x^2/2 + 175,$$

and, in general, all functions of the form

$$x^2/2 + c.$$

Most differential equations are considerably more difficult. The science of solving them consists of hundreds of tricks, each of which will work for only a restricted class of equations.

Consider the equation $f'(x) = 3f(x)$. The series technique is to assume that f can be written as an infinite polynomial

$$f(x) = a_0 + a_1x + a_2x^2 + a_3x^3 + \cdots$$

for some numbers $a_0, a_1, a_2, \ldots$. To discover what these numbers are we use the equation as follows: If $f(x)$ is the polynomial given above, then

$$f'(x) = a_1 + 2a_2x + 3a_3x^2 + \cdots$$

by theorem 10.4, and

$$3f(x) = 3a_0 + 3a_1x + 3a_2x^2 + \cdots.$$

The equation tells us that these two polynomials are equal. We also know that if two polynomials are equal, then the constant terms must be equal, the coefficients of x must be equal, and so on:

$$a_1 = 3a_0$$

$$2a_2 = 3a_1$$

$$3a_3 = 3a_2$$

$$4a_4 = 3a_3$$

$$\vdots$$

Combining equations, we get

$$\left(\underset{\substack{e_i}}{\underbrace{\quad\quad}} \overset{-r \quad\quad c \quad\quad d_i \; b \quad\quad\quad\quad r}{\rule{0pt}{0pt}} \underset{s}{} \right)$$

such that

$$(i - 1)e_i^{i-2} = \frac{c^{i-1} - d_i^{i-1}}{c - d_i}.$$

Then

$$0 \leq \left| F(c) - \frac{f(c) - f(b)}{c - b} \right|$$

$$= \left| \sum_{i=0}^{\infty} ia_ic^{i-1} - \frac{1}{c - b}\left(\sum_{i=0}^{\infty} a_ic^i - \sum_{i=0}^{\infty} a_ib^i \right) \right|$$

$$= \left| \sum_{i=0}^{\infty} ia_ic^{i-1} - \sum_{i=0}^{\infty} ia_id_i^{i-1} \right|$$

$$= \left| \sum_{i=0}^{\infty} ia_i(c^{i-1} - d_i^{i-1}) \right|$$

$$= \left| \sum_{i=0}^{\infty} ia_i(i - 1)(c - d_i)e_i^{i-2} \right|$$

$$\leq \sum_{i=0}^{\infty} \left| i(i - 1)a_i(c - d_i)e_i^{i-2} \right|$$

$$\leq \sum_{i=0}^{\infty} \left| i(i - 1)a_i(c - b)s^{i-2} \right|$$

$$\leq |c - b| \sum_{i=0}^{\infty} \left| i(i - 1)a_is^{i-2} \right|.$$

Since $\sum_{i=0}^{\infty} |i(i - 1)a_is^{i-2}|$ converges by two applications of the previous theorem, it equals some real number t. Then

$$0 \leq \left| F(c) - \frac{f(c) - f(b)}{c - b} \right| \leq |c - b|t$$

and so, if $c - b$ is infinitesimal,

$$F(c) \approx \frac{f(c) - f(b)}{c - b}.$$

Thus $F(c) = f'(c)$. $\square$

$a_0 = a_0$

$a_1 = 3a_0$

$a_2 = (3/2)a_1 = (9/2)a_0$

$a_3 = a_2 = (9/2)a_0$

$a_4 = (3/4)a_3 = (27/8)a_0$

$\vdots$

so $f(x) = a_0 + 3a_0x + \dfrac{9a_0x^2}{2}$

$+ \dfrac{9a_0x^3}{2} + \cdots$

$= a_0(1 + 3x + 9x^2/2 + 9x^3/2$

$+ 27x^4/8 + \cdots).$

This is the solution to the equation. As before, there are many solutions, for a_0 can be any number.

Actually, you might look at the answer a little closer and see that it is

$a_0\left(1 + 3x + \dfrac{(3x)^2}{2} + \dfrac{(3x)^3}{6}\right.$

$\left. + \dfrac{(3x)^4}{24} + \cdots\right) = a_0e^{3x}.$

If we are further given that $f(0) = 3$, we can find a value for a_0 by substituting 0 for x in

$f(x) = a_0 + a_1x + a_2x^2 + a_3x^3 + \cdots,$

obtaining $3 = f(0) = a_0$.

EXERCISES
In each of the following differential equations, find the first five terms of the series for the solution:
1. $f'(x) = x + f(x)$,
$\quad f(0) = 1$.
2. $f'(x) = e^x + f(x)$,
$\quad f(0) = 0$.
3. $f'(x) = x^2 + f(x)$,
$\quad f(1) = 0$. (*Hint:* Assume
$f(x) = a_0 + a_1(x - 1) + a_2(x - 1)^2$
$+ \cdots$ and find $a_0, a_1, a_2, a_3,$ and a_4.)
4. $f''(x) = x + f(x)$,
$\quad f(0) = 1, \; f'(0) = -1$.
5. $f(x) + f'(x) = e^x + f''(x)$,
$\quad f(0) = 2, \; f'(0) = 3$.
6. $f(x) = x + g'(x)$,
$\quad f'(x) = x - g(x)$,
$\quad f(0) = \frac{1}{2}, \; g(0) = \frac{1}{3}$.

THEOREM 10.5. If $f(x) = \sum_{i=0}^{\infty} a_ix^i$ converges in $(-r, r)$, then $F(x) = \sum_{i=0}^{\infty} \dfrac{a_i}{i + 1} x^{i+1}$ converges in $(-r, r)$, and F is a primitive of f.

PROOF: For any x in $(-r, r)$,

$$\sum_{i=0}^{\infty} \left| \frac{a_i}{i + 1} x^{i+1} \right| \leq |x| \sum_{i=0}^{\infty} |a_ix^i|,$$

which converges. $F'(x) = f(x)$ by theorem 10.4; hence, by the Fundamental Theorem of Calculus, F is a primitive of f. $\square$

11

The Topology of the Real Line

In this chapter we present a number of the more important classical theorems of real analysis. These are theorems of fundamental importance that make basic, positive, and definitive statements. They were proved during the formative period by the founders of the field.

Real analysis, the branch of theoretical mathematics extending from the infinitesimal calculus and concerned with real numbers, sets of real numbers, and real-valued functions, was developed largely during the nineteenth century. The pioneering mathematicians whose theorems will appear here include: Karl Weierstrass (1815–1916), Georg Cantor (1845–1918), Richard Dedekind (1831–1916), Emile Borel (1871–1956), Bernard Bolzano (1781–1848), and E. Heine (1821–1881).

Open Sets and Closed Sets

We begin by considering the notion of an open set, a generalization of an open interval that lies at the very foundation of analysis and topology. The key property enjoyed by an open interval (a, b) is that if a real number x lies within it, then all hyperreals infinitely close to x also lie within it.

In this chapter we prove many theorems about sets of *reals*. Frequently, we will be discussing such a set B of reals and will ask whether or not a given nonstandard number h is in B. How can a nonstandard number be in a set of reals? The answer is that when we say "$h \in B$" we mean that the sentence "$h \in B$" is true in ℝ, that ℝ thinks that h is in B. For example, $[0, 1]$ is a set of reals, yet if $\bigcirc > 0$ is an infinitesimal, we still say $\bigcirc \in [0, 1]$.

For example, if $x < h$, $x \approx h$, then $h \in (x, b) \subseteq (a, b)$. This property is the essence of our definition.

DEFINITION. A set B is called *open* iff for any real x, if $x \in B$ and $h \approx x$, then $h \in B$.

We have just noted that open intervals (infinite as well as finite) are open sets. In addition, $\varnothing$ and **R** are clearly open sets. More elaborate open sets can be formed by taking arbitrary unions and finite intersections.

Some examples of sets that are *not* open:
(a) $\{1\}$ — since $1 \in \{1\}$ but $1 + \bigcirc \notin \{1\}$.
(b) $[0, 1]$ — since if $\bigcirc < 0$ is an infinitesimal, then $\bigcirc \notin [0, 1]$.

THEOREM 11.1. The union of any number of open sets is open.

The intersection of any number of open sets might not be open. For example, for each n let

$$A_n = \left(1 - \frac{1}{n}, 1 + \frac{1}{n}\right).$$

Then

$$A = \bigcap_{n<0} A_n = \{1\},$$

not an open set.

On the other hand, what is wrong with this proof that A *is* open? Suppose $r \in A$ and $h \approx r$. Since $r \in A_n$, for each n, and since A_n is open, h must be in A_n also. Thus h is in A_n for all n, so $h \in A$.

The difficulty is that to decide if a real s is in A, we only have to see if s in each A_n, for n finite. But to decide if h is in A, we must see if h is in A_n even for *infinite n*. In this case we can find an infinite N such that $1/N < |h - 1|$, so $h \notin A_N$.

I have no fault with those who teach geometry. That science is the only one which has not produced sects.
Frederick the Great (1712–1786)

Some sets are neither open nor closed, for example, $[0, 1)$. More important, consider Q, the rationals. Using the fact that between any two numbers there is an irrational and a rational, one can show that Q is neither open nor closed. For example, let ⓞ be an irrational between 0 and $1/N$. Then $0 \in Q$ but $0 \approx$ ⓞ $\notin Q$, so Q is not open.

PROOF: Suppose A is the union of open sets, and $x \in A$ is real. By the nature of A, $x \in B$ for some open set $B \subseteq A$. Thus if $h \approx x$, $h \in B \subseteq A$. As x was arbitrary, we have proved that A is open. $\square$

THEOREM 11.2. The intersection of any finite number of open sets is open.

PROOF: Suppose that A_1, A_2, ... , A_n are open sets, and that x is a given real number in $A = A_1 \cap A_2 \cap \ldots \cap A_n$. Then if $h \approx x$, h must also be in A_1, A_2 ... , A_n, as each is open. Thus $h \in A$, and since x was arbitrary, A must be open. $\square$

Corresponding to the key property for open intervals, we have a similar property for closed intervals, namely, if h is a hyperreal in $[a, b]$, and $h \approx x$, a real, then $x \in [a, b]$.

This is simply a restatement of theorem 4.6, since $x = \boxed{h}$. Using this property we define arbitrary closed sets as follows:

DEFINITION. A set B is called *closed* iff whenever $h \in B$ and $h \approx x$, a real, then $x \in B$.

The relationship between open and closed sets is simple.

THEOREM 11.3. A set B is open iff the complement of B, B^c, is closed.

PROOF: Suppose B is open, $h \in B^c$, and $h \approx x$, a real. We must show that $x \in B^c$. This, however, must be the case, for if $x \notin B^c$, then $x \in B$, and thus h would also have to be in B since B is open. Hence $x \in B^c$ and B^c is closed.

Suppose B^c is closed, $x \in B$, a real, and $x \approx h$. We must show that $h \in B$. Again, this must be true, for if $h \in B^c$, then x would have to be in B^c. Hence B is open. $\square$

As with open sets, new closed sets may be formed from old ones by unions and intersections.

THEOREM 11.4. The intersection of any number of closed sets is closed. The union of a finite number of closed sets is closed.

The Topology of the Real Line   107

Theorem 11.4 can be proved in the same way as theorems 11.1 and 11.2, but it can also be proved *from* theorems 11.1, 11.2, and 11.3, and De Morgan's Laws:

$(A \cup B)^c = A^c \cap B^c$

$(A \cap B)^c = A^c \cup B^c.$

Theorem 11.5 is false for open sets. For example, let $A_n = (0, 1/n)$.

The proof is left to the reader.

A simple, yet important theorem concerning closed sets is the following, due to Cantor:

THEOREM 11.5. If $A_1 \supseteq A_2 \supseteq A_3 \supseteq \ldots$ is a sequence of nonempty closed sets, A_1 bounded, each set contained in the one before it, then the intersection of the sets, $\bigcap_{n<\infty} A_n$, is nonempty.

PROOF: Since the sentence, "For all n, $A_n \neq \varnothing$" is true in **R**, it is true in HR. Let $h \in A_N$, for some infinite N. Since $A_N \subseteq A_1$, which is bounded, h lies between two real numbers and is therefore finite. Thus $[h]$ exists. Since $h \in A_n$ for all n, $[h] \in A_n$ for all n. Hence $[h] \in \bigcap_{n<\infty} A_n$. □

A second and very important way to distinguish open and closed sets is by defining certain types of points.

DEFINITION. A real number $b \in B$ is an *interior point of* B iff for *all* $h \approx b$, $h \in B$.

DEFINITION. A real number c is a *limit point of* B iff for *some* $h \in B$, different from c, $h \approx c$.

Note that a limit point c of B need not be in B. For example, a is a limit point of (a, b), but $a \notin (a, b)$.

THEOREM 11.6. A set B is open iff all its points are interior points. A set B is closed iff it contains all its limit points.

EXERCISES
1. Prove theorem 11.4 in the manner of theorems 11.1 and 11.2.
2. Prove theorem 11.6.

EXERCISE
Prove theorem 11.4 using De Morgan's Laws.

DEFINITION. For any set B, the *interior of B*, written B^0, is the set of interior points of B. The *closure of B*, written $\bar{B}$, is the set of all points that are either in B or limit points of B.

Another interesting set is the frontier, or boundary, of a set.

DEFINITION. For any set B, the *frontier* of B is Fr $(B) = \bar{B} \cap \overline{B^c}$.

EXERCISE
Find the frontier of each set in problem 3.

3. For each of the following sets, find the interior and closure:

(a) (0, 1)   (f) the integers
(b) [0, 1]   (g) the rationals
(c) (0, 1]   (h) the irrationals
(d) {4}   (i) **R**
(e) {1, 1/2, 1/3, ...}   (j) ∅

1. Find sets B and C such that
   $\bar{B} \cap \bar{C} \neq \overline{B \cap C}$.
2. Find sets B and C such that
   $B^0 \cup C^0 \neq (B \cup C)^0$.

If you found the frontiers correctly in problem 3, you will know that

$$\mathrm{Fr}(Q) = \mathbf{R}$$

and

$$\varnothing = \mathrm{Fr}(\mathrm{Fr}(Q)) \neq \mathrm{Fr}(Q).$$

Oddly enough, it *is* true that for all A,

$$\mathrm{Fr}(\mathrm{Fr}(\mathrm{Fr}(A))) = \mathrm{Fr}(\mathrm{Fr}(A)).$$

You might try to prove this. (*Hint*: First prove that if B is closed, then $\mathrm{Fr}(\mathrm{Fr}(B)) = \mathrm{Fr}(B)$.)

It is possible to define compact sets for other systems, and in these systems, compact is not necessarily the same as closed and bounded.

4. Prove for all sets B, C:
   (a) $\bar{B} \cup \bar{C} = \overline{B \cup C}$.
   (b) $B^0 \cap C^0 = (B \cap C)^0$.
5. Prove that for all A, $(A^0)^c = \overline{A^c}$.
6. Prove that for all sets A:
   (a) A^0 is open.   (c) $(A^0)^0 = A^0$.
   (b) $\bar{A}$ is closed.   (d) $\overline{(\bar{A})} = \bar{A}$.

Compact Sets

In the real numbers, sets that are both closed and bounded have a number of important properties. We call these sets compact, and they can be defined very neatly.

DEFINITION. A set B is *compact* iff for all $h \in B$, $\boxed{h}$ exists and is also in B.

THEOREM 11.7. For all sets B, B is compact iff B is both closed and bounded.

PROOF: Suppose B is compact. If $h \in B$ and $h \approx x$, x real, then $x = \boxed{h} \in B$. Thus B is closed. Suppose B is not bounded. Then for each $n > 0$, there must be an element $b \in B$ such that $|b| > n$. Let b be such that $b \in B$ and $|b| > N$ for some positive, infinite N. Then $\boxed{h}$ does not exist, a contradiction, and so B is bounded.

Now suppose B is both closed and bounded. If $h \in B$, then the boundedness of B says that $\boxed{h}$ exists, and the fact that B is closed says that $\boxed{h} \in B$. Thus B is compact. $\square$

There are two particularly distinguished theorems concerning compact sets. In addition to their fundamental position in analysis, they are also heavily used in topology.

THE BOLZANO-WEIERSTRASS THEOREM. If B is compact, then every infinite subset of B has a limit point in B.

A shorter proof of this theorem is available to those who have read Chapter 3. Let a be the hyperreal represented by the sequence

$$a_1, a_2, a_3, a_4, \ldots,$$

where these are distinct elements of B. Since all the a_n are different, a is not real, so $a \neq \boxed{a} \cdots$. Since each $a_n \in B$, $a \in B$ by theorem 3.1. Thus $\boxed{a} \in B$ is a limit point of C.

PROOF: Suppose $C \subseteq B$ is infinite. Let a_1, a_2, ... be distinct elements of C. Let N be a positive, infinite integer. By compactness, $\boxed{a_N}$ exists and is in B. Since $a_N \in C$, to show that $\boxed{a_N}$ is a limit point of C, we only have to show that $a_N \neq \boxed{a_N}$. But if $a_N = \boxed{a_N}$ then the sentence

$$\exists m(a_m = \boxed{a_N})$$

would be true in $\mathbf{HR}$ and hence in $\mathbf{R}$. Thus $a_m = \boxed{a_N}$ for some finite m. But the a_i are distinct, contradicting

$a_N = \overline{|a_N|} = a_m$. Thus $a_N \neq \overline{|a_N|}$, and $\overline{|a_N|}$ is a limit point of C lying in B. $\square$

Our second theorem is presented in a slightly weaker version.

The unmodified Heine-Borel Theorem says that if B is the union of *any number* of open sets, then B is contained in the union of only finitely many of them. The distinction between this and the modified version is cloaked in the mysteries of infinite sets, which are beyond the scope of this book. All we will say here is that, surprisingly, not all infinite sets are the same size.

The modified Heine-Borel Theorem can also be proved from theorem 11.5. Since for all n, A_n is open, A_n^c is closed. Let $B_1 = B \cap A_1^c$, $B_2 = B_1 \cap A_2^c, \ldots$. Then $B \supseteq B_1 \supseteq B_2 \supseteq \ldots$, and if B is *not* contained in the union of a finite number of the A_n, then each $B_n \neq \varnothing$. Applying theorem 11.4, let $b \in \bigcap_{n<\infty} B_n$. Then $b \in B$, but b is not in any of the sets A_n, a contradiction.

MODIFIED HEINE-BOREL THEOREM. Suppose B is a compact set, and for all n, A_n is an open set. Then if B is contained in the union of all the sets A_n, B is contained in the union of only a finite number of them.

PROOF: Suppose the theorem is false. Then for each n, there is a point $a_n \in B$ which is not in A_i, for each $i < n$. For some infinite, positive integer N, consider a_N. Since B is compact, $\overline{|a_N|}$ exists and is in B. Since B is contained in the union of all the sets A_n, $\overline{|a_N|} \in A_m$ for some finite m. Since A_m is open and $\overline{|a_N|} \approx a_N$, $a_N \in A_m$. This is a contradiction, however, since a_N is not in A_i, for all $i < N$. This proves the theorem. $\square$

Connected Sets

A connected set is one that cannot be separated by two open sets.

DEFINITION. A set C is *connected* iff it is impossible to find two disjoint, open sets A and B such that $C \subseteq A \cup B$ and $A \cap C \neq \varnothing \neq B \cap C$. If C is not connected, we say it is *disconnected*.

This definition merely puts into symbols what we already understood. Intervals are just sets of the sort: (a, b), $[a, b]$, $(a, b]$, $(-\infty, b)$, $[a, \infty)$, etc.

It is easy to find disconnected sets. For example, $C = \{0, 1\}$, the set with only two points, 0 and 1. This set can be separated by the open sets $(-\frac{1}{2}, \frac{1}{2})$ and $(\frac{1}{2}, 1\frac{1}{2})$. It is not obvious, however, what sets *are* connected. Once again, the answer is easily stated: The connected sets are intervals.

DEFINITION. A set B is an *interval* if whenever $a < x < b$ and $a, b \in B$, then $x \in B$.

To put it another way, intervals have no "holes" in them.

THEOREM 11.8. A set C is connected iff C is an interval.

PROOF: Suppose C is a connected set that fails to be an interval, that is, for some $a < x < b$, $a, b \in C$, $x \notin C$. Then the sets

$$a = (-\infty, x) \quad \text{and} \quad B = (x, \infty)$$

are disjoint, open sets that separate C,

that is, $C \subseteq A \cup B$ (since $x \notin C$), and

$$a \in C \cap A \neq \varnothing \quad \text{and} \quad b \in B \cap C \neq \varnothing.$$

Thus C is disconnected, a contradiction.

Suppose, on the other hand, that C is an interval that is disconnected. Let A, B be two disjoint open sets that separate C, that is, $C \subseteq A \cup B$, and there are points a and b such that $a \in A \cap C$ and $b \in B \cap C$. To get a contradiction, we will define a continuous function on $[a, b]$ for which the Intermediate Value Theorem (p. 47) is false: Since C is an interval, $[a, b]$ is contained in C, and we define for any $x \in [a, b]$

$$f(x) = \begin{cases} 0 & \text{if } x \in A \\ 2 & \text{if } x \in B. \end{cases}$$

We have used the Intermediate Value Theorem (IVT) to prove this. Usually, it is the other way around. The common practice is to prove the IVT using this theorem and theorem 11.12.

f certainly violates the Intermediate Value Theorem, since $f(a) = 0 < 1 < 2 = f(b)$, and for no x in $[a, b]$ does $f(x) = 1$. But f is continuous! Suppose $x \in [a, b]$ is real, and $h \approx x$. Then if $x \in A$, so is h, since A is open. Similarly, if $x \in B$, h is too. Thus regardless of which set x is in—A or B—we see that $f(x) = f(h)$ and so f is continuous. This completes the proof of the theorem. $\square$

An important corollary of this is the Cantor-Dedekind Theorem.

THE CANTOR-DEDEKIND THEOREM. Suppose that there are two nonempty intervals A and B such that $\mathbf{R} = A \cup B$ and $A \cap B = \varnothing$. Suppose also that $a \in A$, $b \in B$, and $a < b$. Then either A has a greatest element or B has a least element.

In these pages we have quoted many philosophers on the nature of mathematics, a subject that few philosophers can resist. Often their views of mathematics are intended merely as illustrations of their larger philosophy and are not especially meaningful to mathematicians. One example of this is Georg Hegel, who used philosophy to show that there could be no satellite between Mars and Jupiter, only months before Ceres was discovered. Hegel considered the derivative the "becoming" of magnitude, while the integral was the "has become."

Hegel's philosophy is one of the more difficult ones. Allegedly he said himself, "Only one man has understood me, and even he has not."

PROOF: Since A and B are intervals and $a < b$, all elements of A are smaller than all elements of B. Suppose A has no greatest element. Then A is open, for if $x \in A$ is a real number, there is another real $y \in A$ such that $x < y$

(otherwise x would be greatest in A). But in this case all hyperreals infinitely close to x are less than y and hence in A. Similarly, if B has no least element, B is open. $\mathbf{R}$, however, is an interval, hence by theorem 11.8, connected. Thus A and B cannot both be open, or they would separate $\mathbf{R}$. This proves the theorem. $\square$

EXERCISES

1. Discover which of the sets in problem 3 of the previous exercise section (p. 108) are compact and which are connected.

2. Prove that if $A_1 \supseteq A_2 \supseteq A_3 \supseteq \ldots$ are all compact and nonempty, then the intersection is nonempty.

3. Find an infinite subset of $(0, 1)$ which has no limit point in $(0, 1)$.

4. For each n, find an open set A_n such that $(0, 1)$ is contained in the union of all the A_n, but not in the union of any finite number of them.

5. Prove that the only sets that are both open and closed are $\mathbf{R}$ and $\emptyset$. (*Hint*: use theorem 11.8.)

Continuous Functions

All of the kinds of sets we have previously described bear a special relationship to continuous functions. To describe these relationships, we will use the following possibly familiar notation:

Notation: For any set A and function f,

$$f(A) = \{f(a) \mid a \in A\},$$

$$f^{-1}(A) = \{a \mid f(a) \in A\}.$$

Our first theorem gives us a new definition of continuity.

THEOREM 11.9. A function f is continuous iff for all open sets A, $f^{-1}(A)$ is open.

PROOF: Suppose f is continuous, and A is open. We must show that $f^{-1}(A)$ is open. Suppose $r \in f^{-1}(A)$ is a real number and $h \approx r$. By definition $f(r) \in A$, and by continuity $f(h) \approx f(r)$. Since A is open, $f(h) \in A$, and so $h \in f^{-1}(A)$. Hence $f^{-1}(A)$ is open.

Now suppose that for all open sets A, $f^{-1}(A)$ is open. Let r be a real and $h \approx r$. We must show that $f(r) \approx f(h)$. But if $f(r) \not\approx f(h)$, then the distance between $f(r)$ and $f(h)$ is greater than some real number $s > 0$, that is, $f(h) \notin (f(r) - s, f(r) + s)$. Let us call the open interval $(f(r) - s, f(r) + s) = A$. Since A is open, $f^{-1}(A)$ is

---

I take space to be absolute.
Isaac Newton

I hold space to be something purely relative as time is.
Gottfried Wilhelm Leibniz

Karl Marx and Friedrich Engels spent some time studying the calculus with the hope of using it to expand their economic theories. In a letter to Marx, Engels once wrote, "…the mathematics of variables whose most important part is the infinitesimal calculus, is in essence nothing other than the application of dialectics to mathematical relations."

Marx gave much consideration to the then-current questions of the foundations of the calculus. He did not approve of infinitesimals.

open. Yet, $r \in f^{-1}(A)$, $r \approx h$, and $h \notin f^{-1}(A)$, a contradiction. $\square$

A number of classical theorems on compact sets are associated with continuous functions. Two of them we essentially proved in Chapter 5. If you examine the proof of theorems 5.1 and 5.2, you will see that we didn't need to start with a closed interval $[a, b]$, but rather could have used any compact set. This gives us:

THEOREM 11.10. If f is continuous on a compact set C, then $f(C)$ is bounded, and f attains a maximum and a minimum on C.

Another succinct result is the following:

THEOREM 11.11. If f is continuous and C is compact, then $f(C)$ is compact.

PROOF: Suppose $h \in f(C)$. We must show that $\boxed{h}$ exists and is in $f(C)$. But if $h \in f(C)$, then for some $h^* \in C$, $f(h^*) = h$. Let $r = \boxed{h^*} \in C$. Then $f(r) \in f(C)$, and by continuity $f(r) \approx f(h^*) = h$. Thus

$$\boxed{h} = f(r) \in f(C). \quad \square$$

A similar theorem is true for connected sets.

THEOREM 11.12. If f is continuous and C is connected, then $f(C)$ is connected.

PROOF: By theorem 11.10, we need only show that $f(C)$ is an interval. Let $a < x < b$ be such that $a, b \in f(C)$. We will prove that $x \in f(C)$. Since $a, b \in f(C)$, there must be $a^*, b^* \in C$ such that $f(a^*) = a$ and $f(b^*) = b$. By theorem 5.3, there is some point c^* between a^* and b^* such that $f(c^*) = x$. Since C is connected, it is an interval, and so $c^* \in C$. Thus

$$f(c^*) = x \in f(C). \quad \square$$

There is a stronger form of continuity that is often of great use in proving theorems involving integrals.

DEFINITION. If A is a given set and f is a given function, then f is said to be *uniformly continuous on A* iff for all $a, b \in A$, if $a \approx b$, then $f(a) \approx f(b)$.

At first glance, this seems identical to continuity, but you will note that for continuity, one of the numbers a or b must be real. For uniform continuity, they

True or False?
For all sets A and continuous functions f,
1. $f(\bar{A}) = \overline{f(A)}$.
2. $f(A^0) = f(A)^0$.
3. $f^{-1}(\bar{A}) = \overline{f^{-1}(A)}$.
4. $f^{-1}(A^0) = f^{-1}(A)^0$.
Answers on page 114.

Theorem 11.12 is a more general form of the Intermediate Value Theorem of Chapter 3. In fact, in view of theorem 11.8, this theorem is easily seen to be the same as the Intermediate Value Theorem.

EXERCISE
Use the Intermediate Value Theorem to prove the (one-dimensional) Brouwer Fixed-Point Theorem: If f is a continuous function from $[0, 1]$ to $[0, 1]$, then there is a point $b \in [0, 1]$ such that $f(b) = b$. (*Hint:* Consider $g(x) = f(x) - x$.)

If you look carefully at the proof of theorem 6.2, you will see that to finish, we had to prove that f was uniformly continuous on $[a, b]$.

may both be nonstandard. Nevertheless, for some sets A, continuity and uniform continuity turn out to be equivalent.

THEOREM 11.13. If A is compact, then f is continuous on A iff f is uniformly continuous on A.

To get some perspective on the power of the hyperreal approach, a traditional proof of theorem 11.13 would require several pages.

PROOF: Clearly, uniform continuity implies continuity. Suppose, however, we are given that f is continuous on A, and $a,b \in A$, $a \approx b$. Simply let $r = \boxed{a} = \boxed{b}$. By compactness, $r \in A$. By continuity, $f(a) \approx f(r) \approx f(b)$. $\square$

EXERCISES

1. Prove that if f is continuous and C is a closed set, then $f^{-1}(C)$ is closed.

2. Find an example of a continuous function f and an open set A such that $f(A)$ is not open.

3. Find an example of a continuous function f on $(0, 1)$ that is unbounded. Find one that is bounded but fails to attain a maximum or minimum.

4. Find a continuous function f and a compact set C such that $f^{-1}(C)$ is not compact.

5. Find a function f and a connected set C such that $f^{-1}(C)$ is not connected.

Examples of functions not uniformly continuous:
(a) $f(x) = 1/x$ on $(0, 1)$: Let $N > 0$ be infinite. Then $1/N \approx 1/(N + 1)$ but

$$f\left(\frac{1}{N}\right) \not\approx f\left(\frac{1}{N + 1}\right).$$

(b) $f(x) = x^2$ on $\mathbf{R}$: $N \approx N + 1/N$ but $f(N) \not\approx f(N + 1/N)$.

Answers to true-false questions:
They are all false.

EXERCISE
Find examples to show they are false. (*Hint*: For 1, let $f(x) = 1/x$ and $A = (1, \infty)$.)

Completeness

$\mathbf{R}$ is complete. We proved this in Chapter 9 when we showed that all Cauchy sequences converge. Another variety of completeness applies to the ordering of $\mathbf{R}$ on a line.

DEFINITION. A number b is an *upper bound* for the set B if $x \leq b$ for all $x \in B$. b is the *least upper bound* for B iff b is an upper bound and there are no smaller upper bounds.

Most mathematicians refer to theorem 11.14 as the completeness of the reals. This is because in many abstract mathematical systems, completeness and the least upper bound property are equivalent.

THEOREM 11.14. $\mathbf{R}$ has the "least upper bound property," that is, every nonempty, bounded subset of $\mathbf{R}$ has a least upper bound.

Another proof of this theorem comes from the Cantor-Dedekind Theorem, another sort of completeness. Given B, let

$$A = \{r \in \mathbf{R} \mid r < b \text{ for some } b \in B\}$$

and let $C = A^c$.
Show that A and C are intervals

PROOF: Suppose a set B has an upper bound, b. For every positive integer n, let b_n be the smallest integer such that b_n/n is an upper bound for B. For some infinite, positive integer N, let

$$r = \boxed{\frac{b_N}{N}}.$$

(exercise). Applying the Cantor-Dedekind Theorem, let $r =$ either the greatest element of A or the least element of C. r then turns out to be the least upper bound.

The least upper bound property can be used to show that every finite hyperreal has a standard part. Simply let $A = \{r \mid r < h, r \text{ real}\}$. It is a good exercise to show that A is a nonempty set of reals with an upper bound. It is another good exercise to then show that the least upper bound of A is the unique real number infinitely close to h.

You may recall completeness from notes on p. 21 and p. 41 where we showed that HR was not complete. Another incomplete system is Q, the rationals. $A = \{r \in Q \mid r^2 < 2\}$ has an upper bound 2, but the least upper bound $\sqrt{2}$ is not in Q.

EXERCISE
Show that $\{\text{\textcircled{$\circ$}} \mid \text{\textcircled{$\circ$}}$ is an infinitesimal$\}$ has no least upper bound in HR.

DEFINITION. For $B \subseteq \mathrm{HR}$. A number b is an *almost upper bound* for B iff for all $x \in B$, either $x \leq b$ or $x \approx b$; b is an *almost least upper bound* for B iff b is an upper bound and if d is any other upper bound, either $b \leq d$ or $b \approx d$.

True or False?
1. If $B \neq \varnothing$ has a real almost upper bound, then it has a least almost upper bound.
2. If $B \neq \varnothing$ has a real upper bound, then B has an almost least upper bound.
3. If $B \neq \varnothing$ has an almost upper bound, then it has an almost least almost upper bound.

This will turn out to be the least upper bound for B. First, suppose that x is any real in B. Then

$$x \leq \frac{b_N}{N},$$

and so

$$x = \boxed{x} \leq \boxed{\frac{b_N}{N}} = r.$$

Thus r is an upper bound for B.

Now suppose s is also an upper bound for B. Because of the way we choose b_N, we know that

$$\frac{b_N - 1}{N}$$

is not an upper bound for B, and so

$$\frac{b_N - 1}{N} < s.$$

But

$$r = \boxed{\frac{b_N}{N}} = \boxed{\frac{b_N}{N} - \frac{1}{N}} \leq \boxed{s} = s.$$

Hence r is the least upper bound of B. $\quad\square$

EXERCISES
1. Define lower bound.
2. Define greatest lower bound.
3. Prove that if $A \subseteq \mathbf{R}$ has a lower bound, then it has a greatest lower bound. (*Hint*: Find the least upper bound of $B = \{-r \mid r \in A\}$.)

True or False?
4. If $A \neq \varnothing$ has a lower bound, then it has a least lower bound.
5. If $A \neq \varnothing$ has an upper bound, then the least upper bound is the greatest lower bound of the set of upper bounds.

12

I have at last become fully satisfied that the language and idea of infinitesimals should be used in most elementary instruction—under all safeguards, of course.
Augustus De Morgan (1806–1871)

In this book we have given definitions of standard objects (continuous functions, derivatives, etc.) in a nonstandard way. In this chapter we will examine the relationship between the nonstandard definitions and the standard ones. With only one exception, this relationship is strict equivalence. The standard definitions are particularly useful to the nonstandard analyst because they can be written in our language L while the nonstandard definitions cannot. Later, with the help of the hyperhyperreal numbers, we will prove some deeper theorems of analysis.

Consider the following standard definition:

EXERCISE
Write this definition in L.

STANDARD DEFINITION. A function f is continuous at $x = a$ iff for all $\varepsilon < 0$, there is a $\delta > 0$ such that $|x - a| < \delta$ implies $|f(x) - f(a)| < \varepsilon$.

The closely held consolation of some rationalizing mathematicians that dy and dx are in fact only infinitely small …is a chimera.
Karl Marx (1818–1883)

THEOREM 12.1 The standard and nonstandard definitions of continuity are equivalent.

PROOF: Suppose f is continuous at $x = a$ by the standard definition, and suppose h is a hyperreal such that $a \approx h$. To show that f is continuous at a by the nonstandard definition, we must show that $f(a) \approx f(h)$. But if $\varepsilon > 0$ is any (real) number, then there is another (real) number $\delta > 0$ such that

$$(*) \quad |x - a| < \delta \to |f(x) - f(a)| < \varepsilon.$$

Since (*) must also be true in HR, and since δ is real, $|h - a| < \delta$ and so $|f(h) - f(a)| < \varepsilon$.

EXERCISE
Prove that the sum of two continuous functions is continuous using the standard definition.
(*Hint*: Given f, g continuous at a, $h(x) = f(x) + g(x)$, and given $\varepsilon > 0$, let δ_1, δ_2 be such that

$$0 < |x-a| < \delta_1 \to |f(x) - g(a)| < \varepsilon/2$$

$$0 < |x-a| < \delta_2 \to |f(x) - g(a)| < \varepsilon/2$$

and then show that if δ is the smaller of the two δ_1, δ_2, it satisfies the standard definition.)

This is true for every real $\varepsilon > 0$ and so $|f(h) - f(a)|$ is either infinitesimal or zero. Hence $f(h) \approx f(a)$.

The other way is even easier: Suppose the nonstandard definition holds and we are given $\varepsilon > 0$. We must show that the sentence "there exists a $\delta > 0$ such that $|x - a| < \delta$ implies $|f(x) - f(a)| < \varepsilon$" is true in **R**. Naturally, it suffices to show this sentence true in HR.

But in HR, there *does* exist such a δ, namely, any positive infinitesimal $\circledcirc$. For if $|x - a| < \circledcirc$, then $x \approx a$ and thus $f(x) \approx f(a)$; hence $|f(x) - f(a)| < \varepsilon$. $\square$

All of the various proofs of equivalence follow the previous pattern with little change. For this reason the proofs will generally be left to the reader.

EXERCISE
Write this definition in L.

STANDARD DEFINITION. The derivative of a function f at a is d iff for all $\varepsilon > 0$, there is a $\delta > 0$ such that

$$0 < |\Delta x| < \delta$$

implies

$$\left| \frac{f(a + \Delta x) - f(a)}{\Delta x} - d \right| < \varepsilon.$$

Logicians have found it sometimes useful to analyze the complexity of a sentence. They start by writing the sentence so that all the quantifiers are in front. For example,

$$\forall x \, \exists y (x + 2 = 7 \wedge y = 4).$$

This is called *prenex normal form*. The "complexity" of the sentence is then measured by the number of changes of quantifiers. For example, the sentence above has a

$$\forall$$

followed by a

$$\exists$$

and is thereby considered more complicated than

$$\forall x (x + x = 4),$$

$$\forall x \forall y (x \neq y),$$

or even

$$\exists x \, \exists y \, \exists z (x^2 + y^2 = z^2).$$

This approach, of course, is not perfect because it doesn't pay any attention to that part of the sentence after the quantifiers, but it is useful and reasonable as well. For example, it takes only a few seconds to understand and evaluate the sentences

$$\exists x \, \exists y (x = 5 + y)$$

and

$$\forall x \, \forall y (x = 5 + y).$$

THEOREM 12.2. The standard and nonstandard definitions of derivative are equivalent.

PROOF: Exercise.

The Integral

The integral we have defined exists and is equal to the standard Riemann integral whenever the latter is defined, but the reverse is not true. Our integral is defined for many functions not Riemann-integrable.

The Riemann integral itself is defined in a variety of equivalent ways. The following definitions are reasonably common:

DEFINITION. For any interval $[a, b]$, a *partition* ρ of $[a, b]$ is a finite set of points in the interval which includes a and b:

$$\rho = \{x_0, x_1, \ldots, x_n\}$$

If f is any bounded function on $[a, b]$, and $0 < i \leq n$, let M_i be the least upper bound of $\{f(x) \,|\, x_{i-1} \leq x \leq x_i\}$ and let m_i be the greatest lower bound of $\{f(x) \,|\, x_{i-1} \leq x \leq x_i\}$. The *upper Riemann sum of f with respect to ρ* is

$$U_\rho = \sum_{i=1}^{n} M_i(x_i - x_{i-1}).$$

But the sentences

$$\forall x \, \exists y (x = 5 + y)$$

and

$$\exists x \, \forall y (x = 5 + y)$$

take a little longer to figure out.

Now that we have introduced this idea of complexity, we can use it to compare standard and nonstandard definitions. For continuity the standard definition reads

$$\forall \varepsilon \, \exists \delta \, \forall x (. \, . \, .),$$

but the nonstandard definition reads

$$\forall \odot (\, . \, . \, .).$$

The difference in complexity is clear.

The use of infinitesimals provides an interesting characterization of a continuously differentiable function, that is, a function whose derivative is continuous.

THEOREM. A function f is continuously differentiable on $[a, b]$ iff

$$f'(h) \approx \frac{f(h + \Delta x) - f(h)}{\Delta x}$$

for all $h \in [a, b]$, real or hyperreal, and all infinitesimals Δx.

SKETCH OF PROOF: If f is continuously differentiable, and h and Δx are given, then by the Mean Value Theorem, there is a q between h and $h + \Delta x$ such that

$$f'(q) = \frac{f(h + \Delta x) - f(h)}{\Delta x}.$$

Since $q \approx h$, $f'(q) \approx f'(h)$.

On the other hand, suppose

$$f'(h) \approx \frac{f(h + \Delta x) - f(h)}{\Delta x}$$

for all $h \in [a, b]$ and infinitesimals Δx, and suppose we are given $h \approx x$.

The *lower Riemann sum of f with respect to ρ* is

$$L_\rho = \sum_{i=1}^{n} m_i (x_i - x_{i-1}).$$

From the illustration below, note that the upper and lower Riemann sums of a positive function are graphically represented by the areas of rectangles above and below the function, respectively.

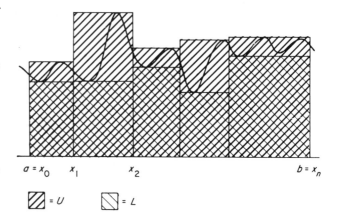

$a = x_0 \quad x_1 \quad\quad x_2 \quad\quad\quad\quad\quad\quad b = x_n$

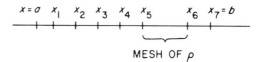

$\boxtimes = U \qquad \boxtimes = L$

DEFINITION. The *mesh* of the partition ρ is the length of the largest subinterval of ρ.

$x = a \quad x_1 \quad x_2 \quad x_3 \quad x_4 \quad x_5 \quad\quad x_6 \quad x_7 = b$

MESH OF ρ

We say a function f is *Riemann-integrable on $[a, b]$* iff there is a number S such that for all $\varepsilon > 0$, there is a $\delta > 0$ such that for all partitions ρ with mesh less than δ,

$$S - \varepsilon < L_\rho \leq S \leq U_\rho < S + \varepsilon.$$

If f is Riemann-integrable, the value of S is called the *Riemann integral* of f.

THEOREM 12.3. If f is Riemann-integrable, then f is integrable and $\int_a^b f(x) \, dx$ equals the Riemann integral of f.

Let $\Delta x = h - x$. Then

$$f'(h) \approx \frac{f(h + (-\Delta x)) - f(h)}{(-\Delta x)}$$

$$= \frac{f(x) - f(x + \Delta x)}{-\Delta x}$$

$$= \frac{f(x + \Delta x) - f(x)}{\Delta x} \approx f'(x).$$

EXERCISE
Prove that the function $f(x) = 17$ is Riemann-integrable on $[0, 1]$.

An example of a Chapter 6-integrable function that is not Riemann-integrable is a little tricky to construct. Leaving out a few details, this is it: First construct a set $S \subseteq (0, 1)$ such that
(1) S is dense in $[0, 1]$ (that is, every open subinterval of $[0, 1]$ contains a point of S), and
(2) if $a, b \in S$, then a/b is irrational.
It is not difficult to construct such a set. Next we define f by

$$f(x) = \begin{cases} 1 & \text{if } x \in S \\ 0 & \text{if } x \notin S. \end{cases}$$

Since S is dense, every partition of $[0, 1]$ will contain points of S (and also points not in S) in each subinterval, so that the lower Riemann sum for $\int_0^1 f(x)dx$ is always 0, and the upper sum is always 1. Thus f is not Riemann-integrable. On the other hand, for any Δx, at most one of the x_i can be in S (where $x_i = i\Delta x$), so that $S_0^1 f(x)\Delta x$ is either 0 or Δx. Thus for any infinitesimal dx,

$$\boxed{\underset{0}{\overset{1}{S}} f(x)dx} = \boxed{0} \text{ or } \boxed{dx} = 0,$$

and so f is Chapter 6-integrable on $[0, 1]$ with $\int_0^1 f(x)dx = 0$.

It is interesting to note that this function f, besides not being Riemann-integrable, is not even Lebesgue-integrable if S is chosen to be a maximal such set.

PROOF: Let S be the Riemann integral of f. For any $\Delta x > 0$, let $\rho_{\Delta x}$ be the partition

$$\rho_{\Delta x} = \{a, a + \Delta x, a + 2\Delta x, \ldots, b\}.$$

The mesh of $\rho_{\Delta x}$ is Δx. You should recognize that this is the partition we used to compute $S_a^b f(x)\,\Delta x$. From the graphic representation below,

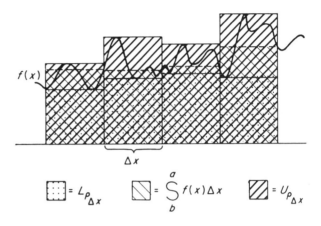

$$\boxed{\vdots} = L_{\rho_{\Delta x}} \qquad \boxed{\diagdown} = \underset{b}{\overset{a}{S}} f(x)\Delta x \qquad \boxed{\diagup} = U_{\rho_{\Delta x}}$$

it is clear that

$$L_{\rho_{\Delta x}} \leq \underset{a}{\overset{b}{S}} f(x)\,\Delta x \leq U_{\rho_{\Delta x}}.$$

Now the definition of Riemann-integrable implies that the sentence "for all $\varepsilon > 0$, there is a $\delta > 0$ such that for all $0 < \Delta x < \delta$,

$$S - \varepsilon < L_{\rho_{\Delta x}} \leq S \leq U_{\rho_{\Delta x}} < S + \varepsilon\text{''}$$

is true in $\mathbb{R}$.

Let ε be a positive infinitesimal $\bigcirc$, and let dx be any positive infinitesimal less than δ. We then have

$$S - \bigcirc < L_{\rho_{dx}} \leq S \leq U_{\rho_{dx}} < S + \bigcirc,$$

and consequently

$$L_{\rho_{dx}} \approx S \approx U_{\rho_{dx}}.$$

Since we also have

$$L_{\rho_{dx}} \leq \underset{a}{\overset{b}{S}} f(x)\,dx \leq U_{\rho_{dx}},$$

we conclude that

$$S \approx \underset{a}{\overset{b}{S}} f(x)\,dx,$$

and so

$$S = \left| \overset{b}{\underset{a}{\mathsf{S}}} f(x)dx \right|$$

since S is real. Thus

$$S = \int_c^b f(x)\,dx. \quad \square$$

In Chapter 9 we discussed the limit of a sequence.

STANDARD DEFINITION. For any sequence $\{a_n\}$, we say $\{a_n\}$ converges to L $(a_n \to L)$ iff for all $\varepsilon > 0$, there is an n such that $m > n$ implies that

$$\left| a_m - L \right| < \varepsilon.$$

Hint: If $a_n \to L$ by the standard definition, and N is infinite, show $|a_N - L| < \varepsilon$ for all real $\varepsilon > 0$.

If $a_n \to L$ by the nonstandard definition, and $\varepsilon > 0$ is real, prove "there exists n such that $m > n$ implies $|a_m - L| < \varepsilon$" is true in ℝ.

THEOREM 12.4. The standard and nonstandard definitions of the convergence of a sequence are equivalent.

PROOF: Exercise.

From Chapter 10 we have a number of new definitions.

STANDARD DEFINITION. A set B is *open* iff for all $b \in B$, there is an $\varepsilon > 0$ such that $|b - x| < \varepsilon$ implies $x \in B$. A set B is *closed* iff every point b with the property "for all $\varepsilon > 0$, there is a point $c \in B$ with $|b - c| < \varepsilon$" is in B.

THEOREM 12.5. The standard and nonstandard definitions of open set are equivalent.

EXERCISE
Compare the complexity of the standard and nonstandard definitions of open set.

PROOF: Suppose B is open by the standard definition, and that $b \in B$ is real with $b \approx h$. We wish to show that $h \in B$. By the standard definition, there is a real $\varepsilon > 0$ such that $|x - b| < \varepsilon$ implies $x \in B$. Since $|h - b| < \varepsilon$, we have that $h \in B$.

EXERCISES
Using the standard definitions, prove the following:
1. The intersection of two open sets is open.
2. A set is open iff its complement is closed.

Suppose now that B is open by the nonstandard definition, and $b \in B$. To show that "there exists an $\varepsilon > 0$ such that $|b - x| < \varepsilon$ implies $x \in B$" is true in ℝ, we need only show that it is true in ℝ. There is, however, such an ε in ℝ, namely, any positive infinitesimal ◎, for if $|b - x| < ◎$ then $b \approx x$. Thus, by the nonstandard definition, $x \in B$. $\quad \square$

EXERCISES
1. Prove the equivalence of the standard and nonstandard definitions of closed set.
2. Give standard definitions of interior point and limit point.

A similar result is true for closed sets. It may be proved directly, or indirectly, by noting that according to either definition, a set is closed if and only if its complement is open.

One last definition from Chapter 10:

There is a direct way, using infinitesimals, of computing second,
third,..., nth derivatives without
computing earlier derivatives. If f
has a second derivative, for example,
then

$f''(x)$

$\approx \dfrac{f(x+2\varDelta x)-2f(x+\varDelta x)+f(x)}{(\varDelta x)^2}$

for any infinitesimal $\varDelta x$. Similarly,
if f has a third derivative,

$f'''(x) \approx \dfrac{[f(x+3\varDelta x)-3(x+2\varDelta x)+3f(x+\varDelta x)-f(x)]}{(\varDelta x)^3}.$

EXERCISES
1. For $f(x) = x^3$, compute directly
$f''(1)$.
2. Guess the formula for $f''''(x)$.

EXERCISE
Prove that the sum of two hyperin-
finitesimals is either hyperinfinitesimal or 0.

STANDARD DEFINITION. A function f is *uniformly con-
tinuous* on a set B iff for all $\varepsilon > 0$, there is a $\delta > 0$ such
that for all $a, b \in B$, $|a - b| < \delta$ implies $|f(a) - f(b)|
< \varepsilon$.

THEOREM 12.6. The standard and nonstandard definitions of uniform continuity are equivalent.

PROOF: Exercise.

The Hyperhyperreal Numbers

The process by which we constructed the hyperreal
numbers from the reals can be repeated to construct
the hyperhyperreal numbers from the hyperreals. The
resulting system will be very useful, particularly because it will contain numbers even smaller than infinitesimals.

DEFINITION. A number $\circledcirc$ is a *hyperinfinitesimal* iff
$0 < |\circledcirc| < h$ for all positive hyperreals h. We say that
a is *hyperinfinitely close to* b ($a \approx b$) iff $a - b$ is either
hyperinfinitesimal or zero.

To describe the power of this new hyperhypersystem,
we must expand our language.

DEFINITION. Let L^{**} be the language L expanded to
include constant symbols for all hyperreal numbers,
function symbols for all hyperreal functions, and relation symbols for all hyperreal relations.

THEOREM 12.7. There exists a number system called the
Hyperhyperreal Numbers (HR) such that
1. HR contains HR.
2. HR contains a hyperinfinitesimal.
3. Every sentence of L^{**} is true in HR iff it is true
in HR.

The proof of this theorem is the exact parallel of
our work in Chapter 3. There is no need to repeat it
here. Further, all theorems of Chapter 4 concerning
infinitesimals and the relation $\approx$ carry over to hyper-
infinitesimals and $\approx$.

As an example of the relationship between R, HR,
and HR, consider the problem of the continuity of a
function f from HR to HR. We again have two definitions, the standard one, which is identical to the one
given at the beginning of this chapter, and the nonstandard one, which is:

DEFINITION. A function f from $\mathbb{HR}$ to $\mathbb{HR}$ is continuous at a hyperreal h iff for all $j \approx h$, $f(j) \approx f(h)$.

The proof that these two definitions are equivalent is virtually identical to the proof of theorem 12.1.

Next consider a function from **R** to **R**. In addition to the two definitions of continuity already given (p. 43 and p. 116), we have another:

DEFINITION. A function f from **R** to **R** is continuous at a real r iff for all $h \approx r$, $f(h) \approx f(r)$.

Once again the proof of equivalence is routine.

Sequences of Functions

DEFINITION. Given a function f, a sequence of functions $\{f_n\}$, and an interval $[a, b]$, we say that the sequence converges to f ($f_n \to f$) on $[a, b]$ iff for all $c \in [a, b]$, $f(c)$ is the limit of the sequence $\{f_n(c)\}$.

Expressing the above definition in nonstandard terms, we have $f_n \to f$ on $[a, b]$ iff for every real x in $[a, b]$ and all infinite integers $N > 0$,

$$f_N(x) \approx f(x).$$

Sequences of functions appear everywhere in analysis and have been studied in hundreds of ways. Their applications to the real world are enormous. However, the abstract theory of sequences is marred by the following fact: *The limit of a sequence of continuous functions may not be continuous.* An example: On the interval $[0, 1]$ consider the functions

$$f_n(x) = x^n.$$

EXERCISE
Prove that if $a \approx b$ and $b \approx c$, then $a \approx c$.

EXERCISES
1. Prove that the nonstandard definition of continuity (p. 43) implies this definition.
2. Prove that this definition implies the standard definition of continuity (p. 116). (*Hint*: Follow the proof of theorem 12.1.)

In Chapter 9 we saw some excellent examples of sequences of functions, namely sequences of the form

$$f_0(x) = f(0)$$

$$f_1(x) = f(0) + f'(0)x$$

$$f_2(x) = f(0) + f'(0)\,x + \frac{f''(0)x^2}{2!}$$

$$\vdots$$

EXERCISES
1. Prove that the nonstandard definition of $f_n \to f$ at right is equivalent to the standard one: "for all x in $[a, b]$ and for all $\varepsilon > 0$, there is an n such that $m > n$ implies

$$|f_m(x) - f(x)| < \varepsilon.\text{"}$$

2. Prove that the standard definition is also equivalent to "for all hyperreal x in $[a, b]$ and all hyperinfinite integers $N > 0$,

$$f_N(x) \approx f(x).\text{"}$$

(*Hint*: Just like 1, except you go between $\mathbb{HR}$ and $\mathbb{HR}$ instead of $\mathbb{HR}$ and **R**.)

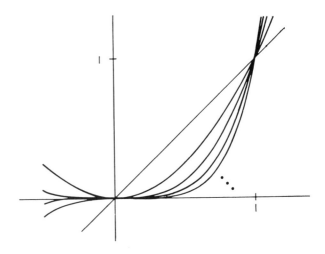

These functions converge to the function

$$f(x) = \begin{cases} 0 & \text{if } x < 1 \\ 1 & \text{if } x = 1, \end{cases}$$

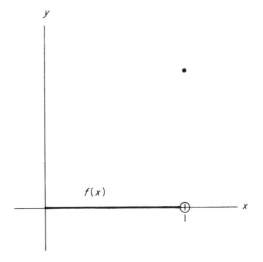

Clearly $f_n(1) \to f(1)$, and although it seems clear that for $a < 1, f_n(a) \to f(a)$, it is more difficult to show. Suppose $0 < a < 1$, and suppose $N > 0$ is an infinite integer. Then a^N must be infinitesimal, for if it were greater than a real $\varepsilon > 0$,

$$\int_a^1 x^N dx$$

would be greater than

$$\int_a^1 \varepsilon \, dx = \varepsilon(1 - a),$$

but

$$\int_a^1 x^N dx = \frac{1}{N + 1} - \frac{a^{N+1}}{N + 1},$$

which is clearly infinitesimal. We also proved that a^N is infinitesimal on p. 88.

A second remedy is to define a weaker sort of continuity. This was done early in the twentieth century by Henri Lebesgue when he defined measurable functions.

Note that in the previous example the convergence is not uniform. Let a be $\sqrt[N]{1/2}$. Then

$$f_N(a) = 1/2 \not\approx 0 = f(a).$$

which is not a continuous function.

One remedy to this situation is to define a stronger sort of convergence, uniform convergence.

DEFINITION. A sequence of functions $\{f_n\}$ converges to a function f *uniformly* on $[a, b]$ iff for every $x \in [a, b]$ and all infinite integers $N > 0$,

$$f_N(x) \approx f(x).$$

Comparing this definition with the previous one, we see that the only difference is that now x may be any number in $[a, b]$, real or nonstandard. This tiny change is exactly what is needed.

THEOREM 12.8. Given a sequence of continuous functions $\{f_n\}$, if $f_n \to f$ uniformly, then f is continuous.

PROOF: The proof is a play on our different definitions of continuity.

Let $x \in [a, b]$, and let h be any hyperhyperreal, $h \approx x$. By definition of continuity (p. 122), we merely must show that

$$f(h) \approx f(x).$$

By uniform convergence we have

$$f(x) \approx f_N(x) \quad \text{and} \quad f_N(h) \approx f(h).$$

The standard definition of uniform convergence is "$f_n \to f$ uniformly on $[a, b]$ iff for all $\varepsilon > 0$, there is an integer $n > 0$ such that for all $m > n$ and all x in $[a, b]$,

$$|f_m(x) - f(x)| < \varepsilon."$$

EXERCISE
Prove that the standard definition is equivalent to the nonstandard definition.

Once again, the uniformity is crucial. Let f_n be the function

$$f_n(x) = \begin{cases} 2n^2x & 0 \le x \le 1/2n \\ 2n - 2n^2x & 1/2n \le x \le 1/n \\ 0 & 1/n \le x \le 1. \end{cases}$$

The sequence $\{f_n\}$ converges to $f(x) = 0$, but

$$\int_0^1 f_n(x)\, dx = 1/2$$

for all n. Thus

$$\int_0^1 f_n(x)\, dx \nrightarrow \int_0^1 f(x)\, dx = 0.$$

EXERCISES
1. Verify these facts.
2. Verify that the hypothesis of a finite interval is necessary in theorem 12.9 as follows: For each n, let

$$f_n(x) = \begin{cases} 0 & \text{if } x < n, x > 2n \\ 1/n & \text{if } n \le x \le 2n. \end{cases}$$

Show that $f_n \to 0$ uniformly on $[0, \infty)$, but that

$$\int_0^\infty f_n\, dx \nrightarrow \int_0^\infty 0\, dx.$$

A STANDARD DEFINITION. A sequence $\{f_n\}$ of continuous functions is *equicontinuous* iff for all a, and $\varepsilon > 0$, there is a $\delta > 0$ such that

$$0 < |x - a| < \delta \to |f_n(x) - f_n(a)| < \varepsilon$$

What about $f_N(x)$ and $f_N(h)$? In **R**, the sentence "for all n, f_n is continuous" is true. Furthermore, using the *standard* definition of continuity, the sentence can be written in L. Thus it is true in **HR**, and so f_N is continuous. Converting back to the nonstandard definition of continuity on **HR**, this implies

$$f_N(x) \approx f_N(h).$$

Putting together the pieces, we have

$$f(x) \approx f_N(x) \approx f_N(h) \approx f(h).$$

Thus

$$f(x) \approx f(h). \quad \square$$

Still better theorems are a consequence of uniform convergence.

THEOREM 12.9. If the functions $\{f_n\}$ converge uniformly to the function f on $[a, b]$, and if each f_n is integrable on $[a, b]$, then f is integrable on $[a, b]$, and

$$\int_a^b f_n(x)\, dx \to \int_a^b f(x)\, dx.$$

PROOF: For finite n, let max (n) be the least upper bound of

$$\{|f(x) - f_n(x)|,\ x \in [a, b]\}.$$

Then for finite Δx, we have

$$\left| \underset{a}{\overset{b}{\mathsf{S}}} f_n(x)\, \Delta x - \underset{a}{\overset{b}{\mathsf{S}}} f(x)\, \Delta x \right| \le (b -)\, a \max(n).$$

Thus the same is true for an infinitesimal $dx > 0$:

$$\left| \underset{a}{\overset{b}{\mathsf{S}}} f_n(x)\, dx - \underset{a}{\overset{b}{\mathsf{S}}} f(x)\, dx \right| \le (b - a)\max(n).$$

Taking standard parts,

$$\left\| \int_a^b f_n(x)\, dx - \underset{a}{\overset{b}{\mathsf{S}}} f(x)\, dx \right\| \le (b - a)\max(n).$$

Then letting n be infinite,

$$\left\| \int_a^b f_N(x)\, dx - \underset{a}{\overset{b}{\mathsf{S}}} f(x)\, dx \right\| \le (b - a)\max(N).$$

for each n (that is, for a given a, ε real, there is one δ that holds for all n).

A NONSTANDARD DEFINITION. A sequence $\{f_n\}$ of continuous functions is *equicontinuous* iff for all a, x, $f_n(a) \approx f_n(x)$ for all n, finite and infinite.

EXERCISES

1. Prove that the nonstandard definition of equicontinuity implies the standard definition. (*Hint*: For a given a, ε real, prove the statement

"$\exists \delta\ \forall x\ \forall n...$"

in HR.)

2. Prove that the standard definition of equicontinuity implies the non-standard definition. (*Hint*: Given b, real, $x \approx b$, show that

$$|f_N(b) - f_N(x)| < \varepsilon$$

for all real $\varepsilon < 0$.)

3. Prove, using each definition, that the sequence of functions $\{x^n\}$ is not equicontinuous.

4. Find an equicontinuous sequence of functions that does not converge to any function.

5. Prove the following theorem:

THEOREM. If f_n is an equicontinuous sequence converging pointwise to a function f, then f is continuous. (*Hint*: Let $N > 0$ be hyperinfinite, then use exer. 2, p. 122, to show

$$f(x) \approx f_N(x) \approx f_N(h) \approx f_N(h)$$

for $x \approx h$.)

But the uniform convergence of f_n implies that $f(x) - f_N(x)$ is always infinitesimal, and so max (N) is infinitesimal. Thus

$$\int_a^b f_N(x)\,dx \approx \boxed{\mathop{S}_a^b f(x)\,dx}$$

and so $\int_a^b f(x)\,dx$ exists and

$$\int_a^b f_n(x)\,dx \to \int_a^b f(x)\,dx. \quad \square$$

Finally, as a corollary to this, we have:

THEOREM 12.10. If the differentiable functions f_n converge to f on $[a, b]$, and if the functions f_n' converge uniformly to g on $[a, b]$, then f is differentiable and

$$f' = g.$$

PROOF: By theorem 12.9. for all c, $d \in [a, b]$,

$$\int_c^d g(x)\,dx \approx \int_c^d f_N'(x)\,dx.$$

By the Fundamental Theorem of Calculus,

$$\int_c^d f_N'(x)\,dx = f_N(d) - f_N(c).$$

Since $\int_c^d g(x)\,dx$ and $f(d) - f(c)$ are both real and

$$\int_c^d g(x)\,dx \approx f_N(d) - f_N(c) \approx f(d) - f(c),$$

we must have

$$\int_c^d g(x) = f(d) - f(c).$$

Hence f is a primitive of g, and so $f' = g.$ $\quad \square$

Appendix A
Defining Quasi-big Sets

Our task here is to define which subsets of N, the natural numbers, are to be called "quasi-big," so that the following four properties hold:

1. No finite set is quasi-big.

The collection of quasi-big sets is called an *ultrafilter* on N. Ultrafilters can be constructed on many sets, and their properties have been relentlessly studied.

2. If A and B are quasi-big, then $A \cap B$ is quasi-big.
3. If A is quasi-big and $A \subseteq C$, then C is quasi-big.
4. If A isn't quasi-big, then A^c (the complement of A) is quasi-big.

For those readers who are familiar with it, the existence of the quasi-big sets, or, in fact, of an ultrafilter on any set, is a straightforward application of Zorn's Lemma.

Our first step is to consider the so-called cofinite sets, the sets whose complements are finite. These sets must be quasi-big by properties (1) and (4) above. But are they *all* the quasi-big sets? No, since, for example, property (4) implies that either the set of even integers or the set of odd integers must be quasi-big, yet neither is cofinite.

The cofinite sets clearly satisfy (1). For (2), suppose A and B are cofinite. Then A^c and B^c are finite, so $A^c \cup B^c$ is finite, hence $(A^c \cup B^c)^c$ is cofinite; but, by De Morgan's law,

$$(A^c \cup B^c)^c = A \cap B.$$

For (3), if A is cofinite and $A \subseteq B$, then $B^c \subseteq A^c$. Since A^c is finite, so is B^c, and hence B is cofinite.

Our plan is to define the quasi-big sets slowly, starting with the cofinite sets. At each stage we will define new sets to be quasi-big, and at each stage we will make sure that the quasi-big sets so far defined still satisfy properties (1) and (2).

We begin by listing all subsets of N:

A_1

It turns out that this is no ordinary list. There are *lots* of subsets of N, and, as a result, the list will be longer than many infinite lists. The listing described here and the subsequent proof can be achieved regardless of the length.

A_2

A_3

A_4

$\vdots$

The technology used here is called "transfinite induction" and is no more mysterious than ordinary induction.

At each stage we will consider one such set, say A, and will make either A or A^c quasi-big. In this way when we have fully gone through the list of subsets of N, the quasi-big sets will have to satisfy property (4).

Stage 1. We check to see if A_1 intersects all quasi-big sets so far defined (in this case, the cofinite sets), that is, we check if for all quasi-big B, $A_1 \cap B \neq \varnothing$. Either A_1 or A_1^c must have this property, for if B and C are quasi-big sets such that $A_1 \cap B = \varnothing$ and $A_1^c \cap C = \varnothing$, then we would have $B \cap C = \varnothing$, violating properties (1) and (2), a contradiction.

These proofs are much easier to comprehend with Venn diagrams. In this case, if

$$A_1 \cap B = \varnothing = A_1^c \cap C,$$

we have:

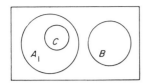

Hence $B \cap C = \varnothing$.

There is something arbitrary about our construction of quasi-big sets. Two different lists of subsets of N might yield two different collections of quasi-big sets. For example, if the set of even integers

$$A = \{2, 4, 6, 8, \ldots\}$$

appears first in the list, it will be made quasi-big by our construction. If, on the other hand, the odd numbers.

$$B = \{1, 3, 5, 7, \ldots\}$$

appears first, it will be made quasi-big and A will not.

Different collections of quasi-big sets may in turn yield different hyperreal number systems, but as we mentioned in Chapter 3, for the purposes of doing calculus, any one hyperreal number system is as good as another.

Whichever of the two sets, A_1 or A_1^c, intersects all quasi-big sets thus far defined we define to be quasi-big. (If both sets intersect all quasi-big sets, we simply define A_1 to be quasi-big.) In addition, if we denote by "A" our quasi-big set just defined, then we also define $A \cap B$ to be quasi-big for each of our old quasi-big sets B.

In this way we enlarge our collection of those sets we're calling quasi-big. Does the enlarged collection still satisfy (1) and (2)? Yes.

1. Let $A \cap B$ be any new quasi-big set. Could it be finite? No, for if it were, then $(A \cap B)^c$ would be cofinite (hence an old quasi-big set), and so $(A \cap B)^c \cap B$ would have to be an old quasi-big set (as the old quasi-big sets satisfy (2)); thus $[(A \cap B)^c \cap B] \cap A \neq \varnothing$ by our choice of A. But $[(A \cap B)^c \cap B] \cap A = \varnothing$. This contradicts the fact that the old quasi-big sets satisfy (1) and so $A \cap B$ must be infinite.

2. For any two new quasi-big sets $A \cap B$ and $A \cap C$, we have that the intersection $(A \cap B) \cap (A \cap C) = A \cap (B \cap C)$ is also a new quasi-big set. Similarly, the intersection of a new and an old quasi-big set is quasi-big. We have thus shown that the enlarged collection of quasi-big sets still satisfies (1) and (2).

Stage 2. At stage 2 we proceed as before except that this time we consider A_2.

Continuing in this way we handle all stages exactly as the first: at stage α we consider A_α and make either A_α or A_α^c quasi-big.

At the end of this process, we have the quasi-big sets. It is clear from the process that these sets satisfy (1) and (2), since they have all along, and of course they satisfy (4) by construction. As we have seen from the left column of page 28, satisfying properties (1), (2), and (4) implies satisfying property (3), and our construction is done.

Appendix B
The Proof of Theorem 3.1

THEOREM 3.1. If G is any formula of L^*, then G is true in HR iff $\{n | G_n$ is true in $\mathbf{R}\}$ is quasi-big.

To prove this theorem, we will have to have a better idea of what we mean by a formula of L^*. Our proof will be intimately based on the structure of such formulas. The first step toward this is to define those collections of symbols that represent elements. These collections, which we call *terms*, either describe a specific element (constants), or any element (variables), or they can be a combination of these.

It is a characteristic of logic that its ideas are simpler than the definitions that describe them. Keep this in mind as you read this section. Terms, for example, are exactly those symbols or combinations of symbols that represent individual elements.

DEFINITION. A *term* of a language is either
1. a constant,
2. a variable, or
3. $f(t_1, t_2, t_3,..., t_n)$ where f is an n-place function symbol, and $t_1, t_2, ..., t_n$ are terms.

For example, $\pi^{\sin(x_7-8)}$ is a term, for x_7 and 8 are terms by (2) and (1), and since $\pi^{\sin(x-y)}$ is a function of two variables, $\pi^{\sin(x_7-8)}$ is a term by (3).

These terms, along with relations, connectives, and quantifiers are the building blocks of the language. We now formally define formulas.

DEFINITION. A *formula* of a language is either
1. $R(t_1, t_2, ..., t_n)$, where R is an n-place relation symbol and $t_1, t_2,..., t_n$ are terms,
2. $(F \vee G)$, where F and G are formulas,
3. $(F \wedge G)$, where F and G are formulas,
4. $(F \rightarrow G)$, where F and G are formulas,
5. $\sim F$, where F is a formula,
6. $\exists x_n F(x_n)$, where $F(x_n)$ is a formula in which x_n appears, or
7. $\forall x_n F(x_n)$, where $F(x_n)$ is a formula in which x_n appears.

For example,
$$(x_3 = 2\tfrac{1}{2}) \rightarrow \exists x_2((x_2 < x_3) \vee I(\sin x_4))$$
is a formula. To see this we note that $x_3, 2\tfrac{1}{2}, x_2, \sin x_4$ are terms, and so
$$(x_3 = 2\tfrac{1}{2}), (x_2 < x_3), I(\sin x_4)$$
are formulas by (1) (where $I(...)$ is the relation "— is an integer"). Then
$$(x_2 < x_3) \vee I(\sin x_4)$$
is a formula by (2),
$$\exists x_2((x_2 < x_3) \vee I(\sin x_4))$$
is a formula by (6), and
$$(x_3 = 2\tfrac{1}{2}) \rightarrow \exists x_2((x_2 < x_3) \vee I(\sin x_4))$$
is a formula by (4).

Once again, the idea is simple. Formulas are just collections of symbols built up from relations.

PROOF: Now to prove the theorem. Suppose the theorem is false. Let us suppose there is a formula G of L^* such that either G is true in HR but $\{n | G_n$ is true in $\mathbf{R}\}$ is not quasi-big, or vice versa, that G is false in HR but $\{n | G_n$ is true in $\mathbf{R}\}$ is quasi-big. Let us also suppose that G is the *shortest possible such formula*. We will show successively that G cannot be a formula of type (1), (2), (3), (4), (5), (6), or (7). Since every formula must be of one of these types, we will have found a contradiction, and the theorem will be proved.

G is not of type (1). By the construction of the

Intuitively, this proof is a type of induction. From the top down, we are showing that if a certain formula A is a counterexample to the theorem, then there is a shorter counterexample, and hence an even shorter counterexample, and so on, an impossibility. Viewed from the bottom up, we are proving that formulas that satisfy the theorem when put together by any of the rules (1)–(7) form formulas that still satisfy the theorem.

hyperreals on page 29, the theorem must hold for all sentences of this basic type.

G is not of type (2). Suppose G is the formula $F \vee K$. Since F and K are each shorter formulas than G, the theorem is true for each of them. We will use this in establishing the theorem for G. Suppose that G is true in HR. Then either F or K must be true in HR, say F. By the theorem applied to F, as F is true in HR, $A = \{n | F_n$ is true in $\mathbf{R}\}$ is quasi-big. Let $B = \{n | G_n$ is true in $\mathbf{R}\}$. Noting that for all n, G_n is just $F_n \vee K_n$, we see that $A \subseteq B$. Thus B must also be quasi-big. To prove the converse, suppose that $B = \{n | G_n$ is true in $\mathbf{R}\}$ is quasi-big. Let $A = \{n | F_n$ is true in $\mathbf{R}\}$ and $C = \{n | K_n$ is true in $\mathbf{R}\}$. Clearly, $A \cap C = B$. This implies that at least one of the two, A or C, must be quasi-big. For if not, then A^c and C^c must both be quasi-big, and then $A^c \cap C^c \cap B$ would be quasi-big. But $A^c \cap C^c \cap B = \varnothing$, and this is a contradiction. Suppose, then, that C is quasi-big. Then applying the theorem to K, K must be true in HR. Hence G must be true in HR. (Similarly, if A is quasi-big, then F is true in HR, and so G must be true in HR.) In either case, G must be true in HR, and so the theorem is true for G. This proves that G cannot be of type (2).

Similarly, G cannot be of types (3), (4), or (5).

G cannot be of type (6). Suppose G is the formula $\exists x_k F(x_k)$. Suppose G is true in HR. Then there is a hyperreal j such that $F(j)$ is true in HR. Since $F(j)$ is shorter than G, the theorem is true for F, so $A = \{n | F(j)_n$ is true in $\mathbf{R}\}$ is quasi-big. But $A \subseteq B = \{n | G_n$ is true in $\mathbf{R}\}$, since G_n is $\exists x_k F_n(x_k)$ and $F(j)_n$ is $F_n(j(n))$. Thus B is quasi-big. Conversely, suppose B is quasi-big. Then for every n in B, $\exists x_k F_n(x_k)$ is true in $\mathbf{R}$, so for every such n let us choose number a_n such that $F_n(a_n)$ is true in $\mathbf{R}$. Let us define a hyperreal j by

$$j(n) = \begin{cases} a_n & \text{if } n \in B \\ 0 & \text{if } n \notin B. \end{cases}$$

Then $B \subseteq C = \{n | F_n(j(n))$ is true in $\mathbf{R}\}$, so C is quasi-big. Since $F(j)$ is shorter than G, the theorem holds for $F(j)$, and so $F(j)$ must be true in HR. Thus G must be true in HR. As the theorem is true for G, G cannot be of type (6).

Similarly, G cannot be of type (7).

This concludes the proof of theorem 3.1. $\square$

EXERCISES

1. Prove that G cannot be of type (3).
2. Prove that G cannot be of type (4).
3. Prove that G cannot be of type (5).

.

EXERCISE

Prove that G cannot be of type (7).

Subject Index

Equicontinuous sequence of functions, 124–125
Exponential functions, 82–83
Extreme Value Theorem, 45–47

Finite number, 36
Formula of a language, 128
Fracimals, 93, 100–101
Frontier of a set, 108
Function sequence. *See* Sequence of functions
Function symbol, 20
Fundamental Theorem of Calculus, 77–84

Geometric sequence, 88
Geometric series, 90–93
Green's Theorem, 77

Heine-Borel Theorem, 110
ℝ. *See* Hyperreal numbers
Hyperhyperreal numbers (ℍℝ), 121–122
Hyperinfinitely close, 121
Hyperinfinitesimal, 121
Hyperreal line, 32–41. *See also* Hyperreal numbers
Hyperreal numbers (ℝ), 9, 25–41
  agrees with **R** on sentences from *L*, 30–31
  constructed, 25–37
  contains infinite numbers, 36–37
  contains infinitesimals, 30
  decimal representation, 38–40
  definition, 25
  and functions, 28–29
  and linear order, 33
  and relations, 29

Infinitely close numbers, 36
Infinitely far apart numbers, 39
Infinite number, 36–37
Infinite polynomial, 95–105. *See also* Series
  convergence, 102–103
  differentiation, 103–104
  integration, 105
Infinite sequence. *See* Sequence
Infinite series. *See* Series
Infinitesimal
  definition, 34
  history, 9, 32, 46, 112

present in ℝ, 30
Infinite sum. *See* Series
Integrable function, 56
Integral, 52–64. *See also* Area
  definition, 56
  existence for continuous functions, 60
  of sums, 81
Integral calculus, 52–64. *See also* Integral
Integral test, 93
Integration by parts, 81
Integration by substitution, 81
Interior of a set, 108
Interior point, 108
Intermediate Value Theorem, 47–49, 111
Interval, 51, 110–112
Inverse function, 75
Inverse Function Theorem, 75
Inverse trigonometric functions, 75
Irrational numbers, 8, 22

Language, mathematical, 13–24
Language *L*
  definition, 21
  R and ℝ agree on, 30–31
  warning, 25
  weakness, 26
Language *L**, 31
Language *L***, 121
Least upper bound, 114
L'Hôpital's Rule, 70
Limit of a sequence
  arithmetic combinations, 86
  definition, 86
  infinite limits, 87
Limit point, 108
Linear ordering, 33
Logarithms, 82–83
Logic, mathematical, 11, 12

MacLaurin's series, 99
Maximum of a function, 71–72
Mean Value Theorem, 72–73
  consequences, 70–71
  for integrals, 82
Mesh, 118
Minimum of a function, 71–72
Monad, 36

Name Index

Alhazon (Ibn-al-Haitham), 53
Anaxagoras, 32
Apostol, T., 103
Archimedes, 8–10, 52–53
Aristotle, 20

Banach, S., 52
Barrow, Isaac, 79–80
Berkeley, George, 6
Bernoulli, Jakob, 56, 80
Bernoulli, Johannes, 46, 70, 80
Bolzano, Bernard, 106
Boole, George, 21
Borel, Emile, 106

Cantor, Georg, 22, 108
Cardan, Jerome, 7
Carroll, Lewis. *See* Dodgson, Charles
Cauchy, Augustin, 70, 80, 86
Cavalieri, Bonaventura, 53, 61, 80, 81

Dedekind, Richard, 106
De Morgan, Augustus, 37, 116
Descartes, René, 10
Dodgson, Charles, 20

Engels, Friedrich, 112, 113
Euler, Leonhard, 7

Fermat, Pierre de, 53, 70, 72, 81
Frege, Gottlob, 24

Galileo Galilei, 74, 80
Gauss, Karl Friedrich, 78
Gregory de St. Vincent, 79, 82

Halley, Edmund, 6
Hegel, Georg, 111
Heine, E., 106
Hobbes, Thomas, 13, 78

Ibn-al-Haitham, 53

Kepler, Johannes, 72
Kronecker, Leopold, 3

Lagrange, Joseph, 78

Lebesgue, Henri, 52, 123
Leibniz, Gottfried Wilhelm, 10, 13, 14, 20, 35, 46, 67, 71–72, 112
    anticipated mathematical logic, 12, 21
    approach to calculus, 3–4, 9
    controversy with Newton, 4–5
    invention of the calculus, 3–4
L'Hôpital, Marquis de, 70
Łos, J., 31
Lukasiewicz, J., 24

Marx, Karl, 112, 116

Napier, John, 82
Newton, Sir Isaac, 6, 7–8, 10, 11, 46, 80, 112
    controversy with Leibniz, 4–5
    invention of the calculus, 3–4
Nicomachus, 53

Occam, William of, 24

Peano, Guiseppe, 24
Peirce, C. S., 16

Roberval, Gilles Personne de, 81
Robinson, Abraham, viii, 4, 9, 11
Rota, Gian-Carlo, 3
Russell, Bertrand, 15, 24

St. Vincent, Gregory de. *See* Gregory de St. Vincent
Sarasa, Alfons A., 82
Seki Kōwa, 4
Sheffer, H. M., 16
Stevin, Simon, 38

Tarski, Alfred, 23, 52
Taylor, Brook, 98
Torricelli, Evangelista, 79–80, 81

Venn, Robert, 20

Wallis, John, 13, 78, 81